The International Library of Bioethics

Volume 82

The *International Library of Bioethics* – formerly known as the International Library of Ethics, Law and the New Medicine comprises volumes with an international and interdisciplinary focus on foundational and applied issues in bioethics. With this renewal of a successful series we aim to meet the challenge of our time: how to direct biotechnology to human and other living things' ends, how to deal with changed values in the areas of religion, society, and culture, and how to formulate a new way of thinking, a new bioethics.

The *International Library of Bioethics* focuses on the role of bioethics against the background of increasing globalization and interdependency of the world's cultures and governments, with mutual influencing occurring throughout the world in all fields. The series will continue to focus on perennial issues of aging, mental health, preventive medicine, medical research issues, end of life, biolaw, and other areas of bioethics, whilst expanding into other current and future topics.

We welcome book proposals representing the broad interest of this series' interdisciplinary and international focus. We especially encourage proposals addressing aspects of changes in biological and medical research and clinical health care, health policy, medical and biotechnology, and other applied ethical areas involving living things, with an emphasis on those interventions and alterations that force us to re-examine foundational issues.

More information about this series at http://www.springer.com/series/16538

Kristen Jones-Bonofiglio

Health Care Ethics through the Lens of Moral Distress

 Springer

Kristen Jones-Bonofiglio
School of Nursing/Centre
for Health Care Ethics
Lakehead University
Thunder Bay, ON, Canada

ISSN 2662-9186   ISSN 2662-9194   (electronic)
The International Library of Bioethics
ISBN 978-3-030-56155-0   ISBN 978-3-030-56156-7   (eBook)
https://doi.org/10.1007/978-3-030-56156-7

This Springer imprint is published by the registered company Springer Nature Switzerland AG
The registered company address is: Gewerbestrasse 11, 6330 Cham, Switzerland

The year 2020 has been designated as the Year of the Nurse and the Midwife and this coincides with the 200th anniversary of the birth of Florence Nightingale. Therefore, I wish to dedicate this book to all nurses and their ongoing commitment to quality care and ethics in practice.

Preface

This book calls for a re-examination of traditional approaches to foundational discourses in health care ethics through a lens of moral distress. In contemporary western society, there is ongoing attention to patient-centered and family-involved care. However, an additional commitment to the needs of carers will require a new ethic—one that acknowledges and supports the inherent risks and suffering of caregiving work.

This new ethic takes a compassionate approach to all those in the circle of care and at many different levels. Written from a Canadian nursing perspective, this book captures the broad interests of multidisciplinary health care professionals and students who face experiences of moral distress during everyday ethical issues in their work with others. It acknowledges ethical realities in different health care contexts, including acute and non-acute health care settings.

This book approaches complex ethical issues from a relational perspective and deeply explores the meaning and potential value of moral distress experiences for health care providers across time through a connection to both historical and contemporary scholarship on this topic. It challenges myths that moral distress comes from individuals caring too much, lacking professional boundaries, or running fresh out of compassion.

Health and healing come from being together and working toward positive changes. Recognizing and addressing moral distress experiences among health care professionals is an untapped resource for early identification, mitigation, and perhaps even prevention of many complex ethical issues in everyday practice. Innovative practical and tactical opportunities for addressing moral distress experiences and for building capacity for ethics in contemporary practice environments are provided.

Thunder Bay, Canada   Kristen Jones-Bonofiglio

Acknowledgements

Sincere thanks to Dr. Sonja Grover who graciously taught me the value of integrity, tenacity, and perseverance.

Appreciation to Dr. Dennis Cooley, Ms. Cecil Joselin Simon, Ms. Floor Oosting, and Mr. Christopher Wilby for their generous guidance through the process of publication for this book.

Gratitude to Dr. Jaro Kotalik, Founding Director of Lakehead University Centre for Health Care Ethics, for demonstrating unyielding energy and an unshakable belief in possibilities.

Special recognition to my Associated Medical Services (AMS) Phoenix Fellowship family and their bold commitment to the growth and development of courageous and compassionate leaders in health care.

Maybe last but never least, I wholeheartedly acknowledge the love and support of my devoted husband Giovanni (John). Beside every writer is someone who gently whispers, "you can do this," and sometimes bravely dares to say, "shouldn't you be writing something right now?"

Contents

About the Author

Dr. Kristen Jones-Bonofiglio, Ph.D. is a Registered Nurse. She is an Assistant Professor, School of Nursing, and Director, Centre for Health Care Ethics, at Lakehead University in Thunder Bay, Ontario, Canada. She serves as Head of the 1st Canadian Bioethics Unit, as part of the International Network of the UNESCO Chair in Bioethics (Haifa). She is a proud member of the International Academy of Medical Ethics and Public Health (IAMEPH), the Canadian Association Schools of Nursing (CASN)-Mental Health Interest Group, and the Canadian Federation of Mental Health Nurses (CFMHN; and education committee).

She holds graduate degrees in public health (nursing specialization) and educational studies (cognition and learning specialization) with research and expertise in moral distress experiences and everyday ethical issues among health care providers. Her diverse scholarship activities include interests in health and well-being, and innovative pedagogical approaches. Her clinical background involves acute care and correctional nursing practice. For over a decade she has taught undergraduate nursing, graduate nursing, and online interdisciplinary continuing professional development courses. In 2017, she became a Fellow of the Associated Medical Services (AMS) Phoenix Foundation with arts-based research on teaching and learning compassion in nursing. In 2019, she received a student-nominated Contribution to Teaching award from Lakehead University, which recognizes high levels of teaching performance and innovation. She lives in Northwestern Ontario with her husband, John.

Chapter 1
Chasing the Science

Abstract The concept of moral distress was first developed in the mid 1980s by American philosopher Andrew Jameton. Since then, moral distress has been studied in many health care settings and among various disciplines. Some say that it has been studied enough and therefore does not require further inquiry or attention. However, as contemporary health care continues to embrace and privilege task-focused, service-oriented, and business-driven approaches to care, the time to reinforce the moral foundations of health care practice has never been more crucial. As fast-paced changes in technology converge with timely demands for individual human rights and social justice, the ethics of health care professionals can no longer be sustained by sterile principles, vague codes, and overarching theoretical values alone. The moral dimensions of quality health care are held in relationships; between and among care providers, patients, families, and communities. With that in mind, experiences of moral distress can actually be valued as early warning signs that only engaged carers are in position to be fully attuned to. As such, these experiences are an undervalued resource for highlighting multi-level needs for positive and essential changes.

Keywords Moral distress · Ethics · Nurses · Health care professionals · Decisions

1.1 Introduction

Moral distress can be defined as an experience where a moral decision has been made about what to do in an ethically challenging situation, but the desired action cannot be carried out. It may include feeling forced to respond with a different action or being unable to act at all. Further, moral distress is a sign of an individual's attunement to their moral responsibilities and the ethics of practice. It is not a sign of weakness or an inability to face the tough realities of caring work. As such, moral distress is a lens that can reveal the ethical issues present in health care environments and the challenges of responding to these issues.

The topic of moral distress in research, education, and practice has received increasing recognition since it was first introduced in the 1980s (Pauly et al. 2009).

The academic literature reveals the existence and widespread prevalence of moral distress in a variety of health care settings (Zuzelo 2007). Moral distress experiences have been described as multi-dimensional with the potential to impact individual health care professionals (Aiken et al. 2002; Pendry 2007), team functioning (Austin 2007), patient care and patient outcomes (Corley 2002; Gutierrez 2005; Wilkinson 1987/88), health care organizations (Aiken et al. 2000), and ultimately, the health care system as a whole (Clarke et al. 2001; Kälvemark et al. 2004).

1.2   Moral Distress

Moral distress involves the perception that an unacceptable compromise of one's values, commitments, and/or obligations has occurred because a right action cannot be carried out. The experience itself challenges one's sense of moral agency, identity, and/or integrity. Moral distress is different from an ethical issue or a moral dilemma alone, because it is an individual experience related to the contextual factors of each unique situation (Epstein and Delgato 2010). Thus, every ethical issue does not lead to moral distress. Some individuals may experience moral distress related to the same situation. However, where one health care provider may experience moral distress, another (who is also attuned to ethical issues in practice) may not experience it. Individuals' values, beliefs, education, and past experiences will vary and influence moral distress experiences among and between health care providers.

A common factor among moral distress experiences is that when it occurs, it is closely tied to an ethical issue. Using Margaret Somerville's (2000) metaphor, moral distress is the ethical canary (in the coal mine) whose distress indicates toxins in healthcare environments (Austin 2012). Each individual experiencing moral distress believes that an action is ethically right or wrong, based on the information that is available to them at the time. Thus, interventions to address moral distress are closely linked with opportunities for dialogue to provide further information and to offer ethics education and support to increase understanding about the ethical issue(s) at hand.

Across health care settings, there are four main tensions described by moral distress scholars. The first of these tensions involves the fact that most of the studies on moral distress are exclusively with nurses in American critical care units. The second tension is that many studies on moral distress use a standardized moral distress questionnaire in an attempt to measure the frequency and intensity of moral distress experiences, despite the lack of agreement on a single definition of what moral distress actually is (and also, what it is not). The third tension involves a general lack of recognition for the potentially compounding effects of other variables. Finally, there is a fourth tension involving gaps in evidence-informed knowledge about interventions that may address, and perhaps even prevent, moral distress.

Therefore, discussions and debates about moral distress are far from over. Everyday ethical issues, experiences of moral distress, and opportunities for continuing education and training all contribute to the quality of care provided to individuals, families, and communities. Circumstances that limit capacity to make ethical decisions and to meaningfully contribute to quality patient care set the stage for the development of moral distress experiences among health care providers (Benner and Wrubel 1989; Severinsson 2003). It is precisely this, a close proximity to the suffering of patients and families and the relational nature of caring practice that places health care providers in a position to recognize 'dis-ease' within ethical decisions and give voice to morally distressing situations (Peter and Liaschenko 2004). Therefore, moral distress is a phenomenon that deserves immediate and ongoing scholarly attention.

1.3 Defining the Concept

Moral distress is a complex personal experience that may involve physical and/or psychological manifestations in response to an ethical situation. It was American philosopher, Andrew Jameton (1984), who first coined the term moral distress almost 40 years ago in his work with nursing students. He wrote about three distinct categories:

(1) moral uncertainty (i.e., the ethical issue and/or right action is unknown);
(2) moral dilemma (i.e., clear ethical issue with conflicting choices for action); and finally,
(3) moral distress (i.e., clear ethical issue, clear choice for action, but one cannot act).

Jameton described moral distress as a negative experience closely tied to institutional barriers that has an impact on ethical nursing practice. As examples, barriers to ethical action may include time constraints, power imbalances, role limitations due to organizational policy, or lack of supervisory support.

Later, Jameton (1993) identified two subsets of moral distress that he designated as initial moral distress and reactive moral distress. Initial moral distress is defined as the first emotional reaction to one's values colliding with institutional barriers or conflicting with those of others. Reactive moral distress occurs when inaction (or the inability to complete a preferred action) has taken place and negative feelings brew over time. However, this work has been widely criticized for being too narrow of a working definition for moral distress. For example, Mark Repenshek (2009), an American ethicist, contends that Jameton's concept of moral distress fails to take into consideration the moral views of patients. Further, Australian nursing ethics scholars Megan-Jane Johnstone and Alison Hutchinson (2015) maintain that Jameton's theory of moral distress is too subjective and based solely on the assumption that nurses know the right thing to do.

Table 1.1 Publications on moral distress as indexed by PubMed (as of June 20, 2020)

Publication dates	Search: moral distress	Search: moral distress and nursing
1984–1995	432 articles	141 articles
1996–2006	1,067 articles	418 articles
2007–2017	5,917 articles	1,665 articles
2018–2020 (to June 20, 2020)	2,761 articles	899 articles
Total	10,177 articles	3,123 articles

To date, a more clearly articulated definition has not been widely accepted and the original and modified definitions (Jameton 1984, 1993) continue to provide a common ground for many contemporary studies on moral distress in the academic literature. It is likely that the surge in academic writing on moral distress over the last ten years (see Table 1.1) has been in response to the building need among many health care professionals for a critical appraisal of this concept and its dimensions.

As a departure from Jameton's (1993) focus on barriers to ethical action, Swedish scholar Sofia Kälvemark Sporrong et al. (2007) describe moral distress as a process beginning with a personal moral stressor that implies the need to fulfill a professional obligation. They describe experiences of moral distress as manifesting with physical, emotional, cognitive, and behavioural signs and symptoms (see Fig. 1.1). Examples of symptoms and sequela related to moral distress have been documented to include varying degrees of aches and pains (e.g., headaches), nightmares, heart palpitations, digestive problems, feeling isolated or alienated, a sense of grief, self-doubt, self-blame, self-criticism, decreased self-esteem, a sense of powerlessness, self-disappointment, fear, anxiety, depression, despair, anger, guilt, sadness, frustration, silence, hopelessness, decreased job satisfaction, and/or loss of meaning (Corley 2002; Gutierrez 2005; Hamaideh 2013; Kelly 1998; Nathaniel 2006; Pauly et al. 2012; Ramber et al. 2010; Rushton 1992; Sundin-Huard and Fahy 1999; Wilkinson 1987).

Fig. 1.1 Conceptualizing moral distress

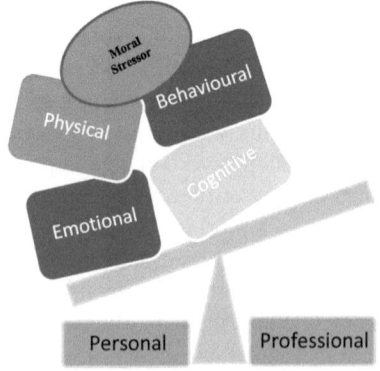

In her work involving interviews with hospital nurses in acute care, Judith Wilkinson (1987) further defined the concept of moral distress as an experience of psychological disequilibrium. As a nurse scholar, she identified a negative impact on almost all her participants' sense of individual wholeness, either personal or professional. Wilkinson's research highlights three significant contributions to expand the dimensions of our understanding of moral distress. These contributions include:

Recap of Concept: Moral Distress

✓ A complex physical, emotional, cognitive, and/or behavioural experience
✓ A negative state of psychological disequilibrium
✓ Going against one's own better judgement
✓ The result of an unmet professional obligation
✓ Linked to internal and external constraints.

(1) additional clinical issues may be linked to moral distress experiences, such as lying and futile treatment;
(2) additional external and internal contraints to ethical practice may occur, such as professional socialization, past failures to enact change, fear, and self-doubt; and,
(3) that moral distress also stems from what one fails to do and/or what one does, against their better judgement (McCarthy and Gastmans 2015).

Further studies and reviews have confirmed the findings of Wilkinson's work on moral distress (such as Huffman and Rittenmeyer 2012). Wilkinson's description of moral distress as a lack of balance (disequilibrium) is an important description for beginning to understand this complex concept and for separating it from other stressors normally associated with nursing practice.

1.4 Additional Theories

There are additional theories that are essential to guide our thinking toward the current understanding of moral distress. The first theory is moral residue, identified by Canadian ethicists Webster and Baylis (2000). George Webster is a clinical ethicist in Manitoba, Canada, with a background in theology. Francois Baylis is a bioethicist, philosopher, and professor in Nova Scotia, Canada. Moral residue is often experienced as pain that is lasting, powerful, and felt deeply and it may remain after values and ideals have been compromised.

In further research exploring moral residue, a concept known as the crescendo effect was described by Epstein and Hamric (2009). Ann Hamric and Elizabeth Epstein are professors of nursing and ethics scholars in the United States of America (USA). Their work attempts to explain how moral distress levels may not return to zero (baseline) after an experience is over because of the presence of moral residue. Rather, repeated exposures to moral distress experiences build a crescendo effect and future responses to morally distressing experiences become more intense and are thus increasingly difficult to process. The crescendo effect model is based on the nursing literature and has not been tested empirically (McCarthy and Gastmans 2015).

The second theory is moral reckoning as described by Alvita Nathaniel (2006), an American nurse and professor. Moral reckoning conceptualizes the process of moral distress and its related sequelae from a grounded theory perspective. According to Nathaniel, moral reckoning involves a three-stage process:

(1) ease (i.e., sense of balance),
(2) resolution (i.e., action), and,
(3) reflection (i.e., emotional impact of consequences).

Nathaniel describes a point of critical juncture, an experience of moral distress, which interrupts the stage of ease. An internal conflict between values and situational forces occurs; however, a decision is required. The second stage of resolution describes the action (i.e., giving up, going along, taking a stand, or bending the rules) that is taken. Finally, in stage three, reflection occurs. Nathaniel describes this stage as the resulting emotional impact on the individual, which can be positive or negative and may be ongoing. The process of moral reckoning as described speaks to the complexities of the phenomenon of moral distress, the various actions that may be considered, and the resulting sequelae that can result upon reflection on the moral distress experience.

The third theory is moral outrage. One outcome of an experience of moral distress can be the development of moral outrage (Pike 1997; Rushton 2013; Wilkinson 1987) in response to violations of integrity and core ethical principles. Moral outrage can be described as justified anger that arises after thoughtful reflection on values such as compassion, empathy, discernment, and humility. Cynda Rushton is an American nursing professor and ethicist. She describes moral outrage as a possible catalyst for necessary action (i.e., compromise, awareness through discussion, refusal to participate, whistle blowing, or exiting from the workplace or situation) (Rushton 2013). However, if an individual's personal judgement of the situation and assessment of timing of a decision do not support necessary action, moral outrage can leave a residue that will not readily heal. Consequences such as apathy and becoming morally mute (Bird 2002) can be the result of unresolved residue from experiences of moral distress.

These three additional theories about moral distress help to develop further understanding of this complex phenomenon. Theory about moral residue and the crescendo effect highlights moral distress as a difficult concept to define and to measure, as one's previous experiences may have unexpected and undetermined cumulative effects. Moral reckoning theory identifies that experiences of moral distress do not necessarily happen at one single, discrete point in time. This theory encourages moral distress researchers to ask questions about actions and non-actions, reflective processes, emotional responses, and behaviours. Finally, moral outrage theory posits that moral distress can have residual effects that can be useful or harmful. Researchers can utilize moral outrage theory for practical application by exploring ways to support positive outcomes from negative moral distress experiences.

1.5 Whose Problem Is It?

There is abundant support for the claim that nurses all over the world experience moral distress in their practice. In addition to many North American studies, research has been conducted about nurses and moral distress across the globe, for example in: **Australia** (Kilcoyne and Dowling 2008), **Brazil** (Barlem et al. 2013; Dalmolin et al. 2012), **China** (Zheng et al. 2015), **Columbia** (Vargas and Concha 2019), **India** (LeBaron et al. 2014), **Iran** (Abdolmaleki et al. 2018; Harorani et al. 2019; Jafari et al. 2019; Khoiee et al. 2008; Shoorideh, Ashktorab, and Yaghmaei 2012; Zabetian et al. 2019), **Ireland** (Dcady and McCarthy 2010), **Israel** (DeKeyser Ganz et al. 2012; Ganz et al. 2015), **Italy** (Karanikola et al. 2013; Lazzari et al. 2019), **Lithuania** (Laurs et al. 2019), **Japan** (Ando and Kawano 2016; Ohnishi et al. 2010), **Jordan** (Hamaideh 2013), **Malawi** (Maluwa et al. 2012), **Netherlands** (Schoot et al. 2006), **Norway** (Forde and Aasland 2008), **Uganda** (Fournier et al. 2007; Harrowing and Mill 2010), **United Republic of Tanzania** (Häggström et al. 2008), **Saudi Arabia** (Rawas 2019), **Sweden** (Silén et al. 2011), and **Taiwan** (Ko et al. 2019).

Although first noted among American nursing students, it would be incorrect to assume that moral distress is exclusively a nursing phenomenon. As Hanna (2004) observed, a predominance of literature on moral distress is based on nurses; however, moral distress has been identified among many other health care professionals and at various levels of leadership. These studies (see Table 1.2) give credence to the need to address moral distress experiences for all members of multidisciplinary health care teams, including students in these professions.

1.6 Exploring Moral Distress

Certain root causes of moral distress were first identified in the literature among acute care nurses but remain as common findings in studies across various health care settings and cultures. These include:

(1) clinical situations, such as end-of-life care decisions (Browning 2013);
(2) internal factors for the individual, such as fear, lack of confidence, perceptions regarding self-efficacy and safety, and self-doubt (Hamric et al. 2006); and,
(3) external factors, such as institutional policies and procedures, fiscal pressures, lack of autonomy, power issues, work environment, ethical climate in the work

Table 1.2 Examples of interdisciplinary studies on moral distress

Cardiovascular implantable electronic device (CIED) industry-employed allied health care professionals (IEAPs)	Mueller et al. (2011)
Childbirth educators	Curl (2009)
Disaster first responders	Gustavsson (2019)
Health care managers	Ganz et al. (2015), Mitton et al. (2010)
Hematopoietic cell transplantation professionals	Neuman et al. (2018)
Interdisciplinary rehabilitation professionals	Mukherjee et al. (2009)
Medical students and residents	Dodek et al. (2019b), Hilliard et al. (2007), Lomis et al. (2009), Wiggleton et al. (2010)
Nursing students (undergraduate)	Sasso et al. (2016); Wojtowicz et al. (2014)
Occupational therapists	Penny et al. (2014)
Pharmacists	Kälvemark Sporrong et al. (2006), Kälvemark Sporrong et al. (2005)
Physicians	Austin et al. (2008), Dodek et al. (2019a), Forde and Aasland (2008), Hamric (2010), Hamric and Blackhall (2007), Lee and Dupree (2008), Lützén et al. (2000), Oliver (2018)
Psychologists	Austin et al. (2005)
Physiotherapists	Carpenter (2010)
Respiratory therapists	Schwenzer and Wang (2006)
Social workers	Mänttäri-van der Kuip (2019), Weinberg (2009)
Unregulated/unlicensed assistive personnel (e.g., health care aides, personal support workers)	Pijl-Zieber et al. (2018)

place, unethical practices in the work setting, poor communication, legal consequences, and lack of administrative support (Hamric 2012; Hamric et al. 2012; Jameton 1993).

Commonly cited situations that hold a high potential for moral distress to occur include: futile or inappropriate treatments, communication issues (i.e., with patients, with families, within interprofessional teams), lack of resources including staffing issues, and incompetence of colleagues resulting in unsafe patient care (Browning 2013; Corley 2002; Epstein and Delgato 2010).

Further, various studies have explored ways to measure the complex phenomenon of moral distress. The first measurement instrument for assessing levels of moral distress was created by Corley et al. (2001). Mary Corley is an American nurse scholar known for her work on moral distress among American critical care nurses (Corley 1995). Called the Moral Distress Scale (MDS), it consists of 38 items and was designed to measure the frequency and intensity of moral distress experiences using a seven-point Likert scale. Corley's research with this measurement tool found moderately high levels of intensity for moral distress among 214 critical care nurses in the United States (Corley et al. 2001). The highest scores correlated with inadequate staffing levels. In a subsequent study of 106 surgical nurses, Corley and her team found a low level of frequency and a moderate level of intensity for moral distress (Corley et al. 2005). This tool was foundational in moral distress research because it was the first attempt to measure the concept. It remains the most widely used quantitative measure for moral distress research (McCarthy and Gastmans 2015).

Further, American nurse scholar Annette Browning's (2013) study of 277 critical care nurses found a moderate to low frequency of moral distress and a high intensity of moral distress. Similar findings, using the MDS, for the frequency of moral distress and its largely negative impact have been reported in a number of subsequent studies (Hamric 2012; Pauly et al. 2012). The translation of findings from the studies noted here indicate that nurses in these studies perceived that a moral distress experience did not happen to them very often, but when it did, the experience was intense.

Many revisions have occurred since the original moral distress scale was created. However, this tool was designed specifically for nurses in critical care settings and was deemed not suitable by the authors for use in other care settings or with other health care providers due to its specific content and context. Approximately a decade later, American scholars Hamric et al. (2012) created the Moral Distress Scale-Revised (MDS-R), for use with a variety of health care professionals in acute care settings. Changes to the MDS included making it shorter, updating expectations for the role of the health care professional, creating more frequently encountered issues for a broader range of care providers, articulating the concepts more clearly, and adding content that the previous scale did not address. The resulting MDS-R has 21 items, uses a five-point Likert scale, and has six parallel versions (for use with nurses, physicians and other providers in adult and pediatric acute care settings). The development and testing of the MDS-R included only physicians and nurses. Further, the MDS-R has undergone psychometric testing and updating to create an MDS-2017 (Whitehead et al. 2018) that aims to capture additional root causes of moral distress. Unfortunately, none of the subsequent adaptations of the scale allow for the appropriate use of this tool in the wide range of community nursing settings because the revisions remain as context and content specific as the original version.

Criticisms of the limitations of the MDS and MDS-R include that they are time consuming and they may actually be measuring moral residue versus a one-time moral distress experience (Wocial and Weaver 2012). Repenshek (2009) issued a warning regarding the wealth of literature that claims to measure moral distress, noting that researchers must be careful and clear as to whether or not they are measuring the contexts in which moral distress experiences occur or measuring the concept of moral distress itself. This is an important caution to consider when reviewing the moral distress literature.

In order to address the shortcomings of the moral distress scale, two American nurse scholars, Lucia Wocial and Michael Weaver (2012) developed and validated the Moral Distress Thermometer (MDT). This screening tool was created for nurses in acute care hospital settings. The MDT uses a visual analogue design and a 0–10 rating scale for self-perceptions of moral distress over a two-week period including the present day. It is described as a quick and easy tool to identify nurses presently at risk for leaving their job due to moral distress and as a means of evaluation for moral distress interventions. With 529 participants from three hospitals in the USA, the MDT results showed statistically significant correlations with the MDS (2009, adult and pediatric versions). According to Wocial and Weaver (2012), the moral distress thermometer requires further testing to determine its reliability.

Other scholars have explored moral distress from a qualitative perspective using interviews (such as Fry et al. 2002; Gunther and Thomas 2006; Hanna 2005; Wilkinson 1987) or using a grounded theory approach (such as Gutierrez 2005). Some have made attempts to more accurately define moral distress by allowing for sections of surveys to have open-ended questions for participants to use their own words to express their moral distress experiences (such as Hamric and Blackhall 2007; Pauly et al. 2009). This is an important next step. Finally, open-ended responses on moral distress questionnaires have been found to indicate more moral distress than statistical results from a survey alone (Wilson et al 2013). These findings have implications for the reliability and validity of studies that claim to measure frequency and intensity of moral distress experiences using number scales. By far, most of the studies on moral distress have a strictly qualitative design with emphasis on nurses in acute care settings in western contexts.

1.7 Interventions

Despite the complexities of moral distress experiences for nurses and other health care providers across many health care sectors, there are means to support coping and practices that achieve a high standard of ethical care. Literature on moral theory and ethics education identifies a philosophical shift towards an emphasis on process versus content alone and an adherence to ethics principles (such as Walker 1993). While this shift is important, there is a need for caution. As Irish nursing scholar Joan McCarthy (2003) points out, there are historical and cultural constraints that must be considered whenever the parameters of ethics are redrawn. The search continues for theories, processes, and educational approaches to health care ethics that are salient, timely, and responsive to everyday practice realities.

The Canadian Nurses Association (2003) notes practice environments to be important and central to decreasing moral distress. Further, leading Canadian experts on moral distress Varcoe et al. (2012) suggest a need to develop a safe moral community for nursing practice based on values that are freely shared to direct our actions. Considering strong evidence in the literature for correlations between moral distress and ethical climate/ethical work environment (such as Corley et al. 2005; Hamric and Blackhall 2007; Hamric et al. 2012; Pauly et al. 2009), a renewed focus on supporting quality practice environments may be one important strategy to consider. Like improving air quality in a coal mine, changes in this area are likely to have a favourable impact on health care professionals by supporting ethical practices in various work environments.

Canadian nurse researcher Wendy Austin et al. (2005) acknowledge that the inter-disciplinary team itself can be a source of moral distress, as well as a resource for resolution of moral distress. Thus, in the development of quality health care practice environments, all members of the team must be considered. It will be within collegial relationships and dialogues that dynamic positive changes first occur. The academic literature suggests enhancing opportunities for shared practice models and inter-professional education (such as Ulrich et al. 2010). Among team members, ethical decision-making can be facilitated using informal or formal approaches (Helft et al. 2009), such as establishing an ethics committee, having pastoral support, or desig-nating ethics consultants (Ulrich et al. 2010). The emphasis should be on supporting important, and ideally interprofessional, dialogues on a regular basis.

The wear and tear of everyday ethical issues and experiences of moral distress in health care environments should not be ignored (Kälvemark Sporrong et al. 2006). Often there is a tendency for ethics education to focus on rational decision-making processes (i.e., algorithms and frameworks) and the use of ethics principles to analyze dilemmas (Walsh 2010). This approach overlooks the contextual and emotional aspects of moral distress, individual's beliefs about what is and what is not ethical, and, specifically, the role of moral agency (Hamric 1999; Rodney 1997). Do nurses feel accountable for participating in ethical decisions? Do nurses feel responsible for the outcome of ethical decisions? From a humble and empathetic perspective, where are the limitations for a client's care and a nurse's obligations? These are important questions because the emotional nature of moral distress can affect nurses personally as well as professionally. As Canadian sociologist Arthur Frank (2004) keenly notes, this shift in thinking about ethics, from decision-making to potentially identity-breaking, highlights that effects ripple out from what I do, to include who I become.

Ongoing education and professional development opportunities offer nurses and other health care providers avenues to find the words to describe their moral distress experiences and explore tangible solutions to very complex ethical dilemmas in practice that are compounded by issues such as lack of communication, poor collab-oration, and power imbalances (Ulrich et al. 2010). Education may be delivered via a variety of modalities, such as workshops, seminars, support groups, or journal clubs (Astrom et al. 1995; Beumer 2008; Burger et al. 1992; Kälvemark Sporrong et al. 2007), with online and in-person options.

Historically, teaching strategies for nursing ethics education have included a principles-based approach (Beauchamp and Childress 2001), ethical decision-making models (Aroskar 1980; Crisham 1985; Curtin 1978), and case study analysis (White et al. 1990). None of these have been found to be particularly helpful for wrestling with complex everyday ethical issues in nursing practice. There are at least three reasons for limited success with these approaches, including that:

(1) they do not generally consider the significance of relationships (e.g., between health care professionals and patients, between professional colleagues);
(2) they do not generally take proper account of the context of each unique situation; and;

(3) they are limited in the scope of understanding of the many factors, from a holistic perspective, that may contribute to an ethical issue (e.g., physical, emotional, mental, social, political, spiritual, cultural, and financial).

Although the above mentioned teaching strategies are not as effective as one would ideally desire, there may be ways to enhance these approaches that could positively contribute to supporting nurses and other health care providers with ethics in practice and moral distress experiences.

An alternate approach to educating nurses about the issue of moral distress is one that emphasizes opportunities to voice real world concerns and to learn practical skills to manage moral distress experiences (Eizenberg et al. 2009). In nursing, the theory-to-practice gap in relation to coping with moral distress in the health care workplace appears to be a wide and deep divide. Although, efforts to support ethics education may successfully bridge that gap for nurses and perhaps other health care providers in various health care settings. In my own graduate research (Jessiman 2008), only a small percentage of the respondents (6%) disagreed that educational initiatives about ethics would be considered particularly helpful to them, both personally and professionally. The opportunity for practice to drive theory and theory to support practice is a necessary iterative process in order to adapt educational strategies to best fit the fast-paced, dynamic environment of contemporary health care systems. And, on the level of each individual health care provider the end goal of education is to encourage self-directed growth (Buchholz and Rosenthal 2001).

1.8 Conclusion

Despite all of the research and academic attention, one could argue that even after forty years of study we are not much further ahead in our understanding of moral distress in health care. In theory, moral distress is a cognitive-emotional experience that results from two co-occurring circumstances: (1) feeling and/or believing that a situation is morally/ethically off track; and, (2) having a sense of responsibility or obligation to act. Although identified as being an individual experience, moral distress does not occur in isolation. It occurs in the context of relationships with real consequences and difficult outcomes that may have a ripple effect. Within these experiences, the dynamics of power, trust, and compassion deserve deeper attention and exploration.

From a strengths-based perspective, it would not be fitting to end this first chapter with such a hopeless feeling about the state of the science on moral distress. What can be valued here? Moral distress experiences occur among engaged care providers who are sensitive to their own ethical and moral compass. This attunement is an asset to the individual carer, their team, their organization, and ultimately to the patients, families, and community that they care for and about. Like a fire that lights the darkness, moral distress can be a valuable resource.

References

Abdolmaleki, M., S. Lakdizaji, A. Ghahramanian, A. Allahbakhshian, and M. Behshid. 2018. Relationship between autonomy and moral distress in emergency nurses. *Indian Journal of Medical Ethics* 6: 1–5. http://doi.org/10.20529/IJME.2018.076.

Aiken, L., S. Clarke, and D. Sloane. 2000. Hospital restructuring: Does it adversely affect care and outcomes? *Journal of Nursing Administration* 30 (10): 457–465.

Aiken, L., S.P. Clarke, D.M. Sloane, J. Sochalski, and J.H. Silber. 2002. Hospital nurse staffing and patient mortality, nurse burnout, and job dissatisfaction. *Journal of American Medical Association* 288 (16): 1987–1993. https://doi.org/10.1001/jama.288.16.1987.

Ando, M., and M. Kawano. 2016. Association between moral distress and job satisfaction of Japanese psychiatric nurses. *Asian/Pacific Island Nursing Journal* 1 (2): 55–60. https://doi.org/10.9741/23736658.1020.

Aroskar, M. 1980. Anatomy of an ethical dilemma: The theory. *American Journal of Nursing* 4: 658–660.

Astrom, G., C. Furåker, and A. Norberg. 1995. Nurses' skills in managing ethically difficult care situations: Interpretations of nurses' narratives. *Journal of Advanced Nursing* 21: 1073–1080.

Austin, W. 2007. Ethics of everyday practice: Healthcare environments as moral communities. *Advances in Nursing Science* 30 (1): 81–88.

Austin, W. 2012. Moral distress and the contemporary plight of health professionals. *HEC Forum* 24 (1): 27–38. https://doi.org/10.1007/s10730-012-9179-8.

Austin, W.J., L. Kagan, M. Rankel, and V. Bergum. 2008. The balancing act: Psychiatrists' experience of moral distress. *Medicine, Health Care and Philosophy* 11: 89–97. https://doi.org/10.1007/s11019-007-9083-1.

Austin, W., G. Lemermeyer, L. Goldberg, V. Bergum, and M. Johnson. 2005. Moral distress in healthcare practice: The situation of nurses. *HEC Forum* 17 (1): 33–48.

Barlem, E., V. Lunardi, G. Lunardi, J. Tomaschewski-Barlem, R. Silveira, R. and G. Dalmolin. 2013. Moral distress in nursing personnel. *Revista Latino-Americana de Enfermagem* 21 (spe): 79–87. http://doi.org/10.1590/S0104-11692013000700011.

Beauchamp, T.L., and J.F. Childress. 2001. *Principles of biomedical ethics*, 5th ed. New York, NY: Oxford University Press.

Benner, P., and J. Wrubel. 1989. *The primacy of caring: Stress and coping in health and illness.* Menlo Park, CA: Addison-Wesley.

Beumer, C. 2008. Innovative solutions: The effect of a workshop on reducing the experience of moral distress in an intensive care unit setting. *Dimensions in Critical Care Nursing* 27 (6): 263–267. https://doi.org/10.1097/01.DCC.0000338871.77658.03.

Bird, F.B. 2002. *The muted conscience: Moral silence and the practice of ethics in business.* Westport, CT: Quorum Books.

Browning, A.M. 2013. Moral distress and psychological empowerment in critical care nurses caring for adults at end of life. *American Journal of Critical Care* 22 (2): 143–151. https://doi.org/10.4037/ajcc2013437.

Buchholz, R.A., and S.B. Rosenthal. 2001. A philosophical framework for case studies. *Journal of Business Ethics* 29: 25–31.

Burger, A.M., J.A. Erlen, and L. Tesone. 1992. Factors influencing ethical decision making in the home setting. *Home Healthcare Nurse* 10 (2): 16–20.

Canadian Nurses Association. 2003. Ethical distress in health care environments. *Ethics in Practice for Registered Nurses* 1–8. Retrieved from https://www.cna-aiic.ca/-/media/cna/page-content/pdf-fr/ethics_pract_ethical_distress_oct_2003_e.pdf.

Carpenter, C. 2010. Moral distress in physical therapy practice. *Physiotherapy Theory and Practice* 26 (2): 69–78. https://doi.org/10.3109/09593980903387878.

Clarke H., H. Spence Laschinger, P. Giovannetti, J. Shamian, D. Thomson, and A. Tourangeau. 2001. Nursing shortages: Workplace environments are essential to the solution. *Hospital Quarterly* 50–58.

Corley, M.C. 1995. Moral distress of critical care nurses. *American Journal of Critical Care* 4 (4): 280–285.

Corley, M.C. 2002. Nurse moral distress: A proposed theory and research agenda. *Nursing Ethics* 9 (6): 636–650. https://doi.org/10.1191/0969733002ne557oa.

Corley, M., R.K. Elswick, M. Gorman, and T. Clor. 2001. Development and evaluation of a moral distress scale. *Journal of Advanced Nursing* 33 (2): 250–256.

Corley, M., P. Minick, R.K. Elswick, and M. Jacobs. 2005. Nurse moral distress and ethical work environment. *Journal of Advanced Nursing* 12 (4): 381–390. https://doi.org/10.1191/096973300 5ne809oa.

Crisham, P. 1985. MORAL: How can I do what's right? *Critical Care Management* 3: 42A–42J.

Curl, M. 2009. Commentary—Chronic moral distress among childbirth educators: Is there a cure? *The Journal of Perinatal Education* 18 (1): 48–50. https://doi.org/10.1624/105812409X396237.

Curtin, L. 1978. A proposed model for critical ethical analysis. *Nursing Forum* 17 (1): 12–17.

Dalmolin, G., V. Lunardi, E. Barlem, and R. Silveira. 2012. Implications of moral distress on nurses and its similarities with burnout. *Texto & Contexto—Enfermagem* 21 (1): 200–208. https://doi.org/10.1590/S0104-07072012000100023.

Deady, R., and J. McCarthy. 2010. A study of the situations, features, and coping mechanisms experienced by Irish psychiatric nurses experiencing moral distress. *Perspectives in Psychiatric Care* 46 (3): 209–220. https://doi.org/10.1111/j.1744-6163.2010.00260.x.

DeKeyser Ganz, F., O. Raanan, R. Khalaila, K. Bennaroch, S. Scherman, M. Bruttin, Z. Sastiel, N. Farkash Fink, and J. Benbenishty. 2012. Moral distress and structural empowerment among a national sample of Israeli intensive care nurses. *Journal of Advanced Nursing* 69 (2): 415–425. https://doi.org/10.1111/j.1365-2648.2012.06020.

Dodek, P.M., E. Cheung, K. Burns, C. Martin, P. Archambault, F. Lauzier, A. Sarti, A. Fox-Robichaud, A. Seely, A. Hamric, C.S. Parshuram, D. Garros, D. Wensley, D. Withington, D.J. Cook, D. Piquette, F. Carnevale, J.G. Boyd, G. Emeriaud, J. Downar, J. Rennick, D.J. Kutsogiannis, M. Dagenais, M. Chasse, P. Fontela, R. Fowler, S.M. Bagshaw, S. Dhanani, S. Murthy, and T. Fujii. 2019a. Moral distress and other wellness measures in Canadian intensive care physicians. *American Journal of Respiratory and Critical Care Medicine* 199: A4301.

Dodek, P.M., A. Culjak, E.O. Cheung, M.M. Hubinette, C. Holmes, B. Schrewe, K. Wisener, and P. Crowell. 2019b. High prevalence of burnout despite low levels of moral distress and high resilience scores in medical students. In *Determinants of Burnout and Wellness among Physicians and Trainees: American Thoracic Society* (Conference Proceeding), vol. C21, A4299–A4299.

Eizenberg, M.M., H.S. Desivilya, and M.J. Hirschfeld. 2009. Moral distress questionnaire for clinical nurses: Instrument development. *Journal of Advanced Nursing* 65 (4): 885–892. https://doi.org/10.1111/j.1365-2648.2008.04945.x.

Epstein, E.G., and S. Delgato. 2010. Understanding and addressing moral distress. *The Online Journal of Issues in Nursing* 15 (3). https://doi.org/10.3912/OJIN.Vol15No03Man01.

Epstein, E.G., and A.B. Hamric. 2009. Moral distress, moral residue, and the crescendo effect. *Journal of Clinical Ethics* 20 (4): 330–342.

Forde, R., and O.G. Aasland. 2008. Moral distress among Norwegian doctors. *Journal of Medical Ethics* 34 (7): 521–525. https://doi.org/10.1136/jme.2007.021246.

Fournier, B., W. Kipp, J. Mill, and M. Walusimbi. 2007. Nursing care of AIDS patients in Uganda. *Journal of Transcultural Nursing* 18 (3): 257–264. https://doi.org/10.1177/1043659607301301.

Frank, A.W. 2004. Ethics in medicine: Ethics as process and practice. *Internal Medicine* 34: 355–357.

Fry, S.T., R.M. Harvey, A.C. Hurley, and B.J. Foley. 2002. Development of a model of moral distress in military nursing. *Nursing Ethics* 9 (4): 373–387. https://doi.org/10.1191/0969733002ne522oa.

Ganz, F.D., N. Wagner, and O. Toren. 2015. Nurse middle manager ethical dilemmas and moral distress. *Nursing Ethics* 22 (1): 43–51. https://doi.org/10.1177/0969733013515490.

Gunther, M., and S.P. Thomas. 2006. Nurses' narratives of unforgettable patient care events. *Journal of Nursing Scholarship* 38 (4): 370–376.

Gustavsson, M. 2019. Moral distress among disaster responders: What is it, and can we do anything about it. A scoping review. Series Report. *PREA Conference. Ethics and Humanitarian Research: Generating Evidence Ethically.* The Fawcett Event Center, Ohio State University Columbus, Ohio, March 25–26, 2019. Presentation. Session 14. Oral Presentation 5. Paper B. https://kb.osu.edu/handle/1811/87658.

Gutierrez, K. 2005. Critical care nurses' perceptions of and responses to moral distress. *Dimensions of Critical Care Nursing* 24 (5): 229–241.

Häggström, E., E. Mbusa, and B. Wadensten. 2008. Nurses' workplace distress and ethical dilemmas in Tanzanian health care. *Nursing Ethics* 15 (4): 478–491. https://doi.org/10.1177/0969733008090519.

Hamaideh S.H. 2013. Moral distress and its correlates among mental health nurses in Jordan. *International Journal of Mental Health Nursing* 23 (1): 33–41. https://doi.org/10.1111/anu.12000.

Hamric, A. 1999. Ethics. The nurse as a moral agent in modern health care. *Nursing Outlook* 47(3):106.

Hamric, A.B. 2010. Moral distress and nurse-physician relationships. *American Medical Association Virtual Mentor* 12 (1): 6–11. https://doi.org/10.1001/virtualmentor.2010.12.1.ccas1-1001.

Hamric, A.B. 2012. Empirical research on moral distress: Issues, challenges, and opportunities. *HEC Forum* 241: 39–49. https://doi.org/10.1007/s10730-012-9177-x.

Hamric, A.B., and L.J. Blackhall. 2007. Nurse-physician perspectives on the care of dying patients in intensive care units: Collaboration, moral distress, and ethical climate. *Critical Care Medicine* 35 (2): 422–429. https://doi.org/10.1097/01.CCM.0000254722.50608.2D.

Hamric, A., C.T. Borchers, and E.G. Epstein. 2012. Development and testing of an instrument to measure moral distress in healthcare professionals. *AJOB Primary Research* 3 (2): 1–9. https://doi.org/10.1080/21507716.2011.652337.

Hamric, A.B., W.S. Davis, and M.D. Childress. 2006. Moral distress in health care professionals. *Pharos* 69 (1): 16–23.

Hanna, D.R. 2004. Moral distress: The state of the science. *Research and Theory for Nursing Practice: An International Journal* 18 (1): 73–93.

Hanna, D.R. 2005. The lived experience of moral distress: Nurses who assisted with elective abortions. *Research and Theory for Nursing Practice* 19 (1): 95–124.

Harorani, M., M. Golitaleb, F. Davodabady, S. Zahedi, M. Houshmand, S.S. Mousavi, and M.S. Yousefi. 2019. Moral distress and self-efficacy among nurses working in critical care unit in Iran: An analytical study. *Journal of Clinical & Diagnostic Research* 13 (11): 6–9.

Harrowing, J., and J. Mill. 2010. Moral distress among Ugandan nurses providing HIV care: A critical ethnography. *International Journal of Nursing Studies* 47: 723–731. https://doi.org/10.1016/j.ijnurstu.2009.11.010.

Helft, P.R., P.D. Bledsoe, M. Hancock, and L.D. Wocial. 2009. Facilitated ethics conversations: A novel program for managing moral distress in bedside nursing staff. *Journal of Nursing Administration's (JONA's) Healthcare Law, Ethics, and Regulation* 11 (1): 27–33. https://doi.org/10.1097/NHL.0b013e31819a787e.

Hilliard, R., C. Harrison, and S. Madden. 2007. Ethical conflicts and moral distress experienced by pediatric residents during their training. *Paediatrics & Child Health* 12 (1): 29–35.

Huffman, D.M., and L. Rittenmeyer. 2012. How professional nurses working in hospital environments experience moral distress: A systematic review. *Critical Care Nursing Clinics of North America* 24 (1): 91–100.

Jafari, M., M. Hosseini, S.S.B. Maddah, H. Khankeh, and A. Ebadi. 2019. Factors behind moral distress among Iranian emergency medical services staff: A qualitative study into their experiences. *Nursing and Midwifery Studies* 8 (4): 195–202. https://doi.org/10.4103/nms.nms_69_18.

Jameton, A. 1984. *Nursing practice: The ethical issues.* Engelwood Cliffs, NJ: Prentice Hall.

Jameton, A. 1993. Dilemmas of moral distress: Moral responsibility and nursing practice. *AWHONN's Clinical Issues in Perinatal and Women's Health Nursing* 4: 542–551.

Jessiman, K. (2008). *Everyday ethics in case management: Experiences of moral distress by professionals in a community health care setting* (unpublished master's thesis). Lakehead University, Thunder Bay, ON, Canada. Retrieved on December 27, 2019 from http://knowledgecommons.lakeheadu.ca:7070/bitstream/handle/2453/3862/JessimanK2008m-1b.pdf?sequence=1.

Johnstone, M.J., and A. Hutchinson. 2015. 'Moral distress'—Time to abandon a flawed nursing construct? *Nursing Ethics* 22 (1): 5–14. https://doi.org/10.1177/0969733013505312.

Kälvemark Sporrong, S., B. Arnetz, M.G. Hansson, P. Westerholm, and A.T. Höglund. 2007. Developing ethical competence in health care organizations. *Nursing Ethics* 14 (6): 825–837. https://doi.org/10.1177/0969733007082142.

Kälvemark Sporrong, S.K., A.T. Hoglund, and B. Arnetz. 2006. Measuring moral distress in pharmacy and clinical practice. *Nursing Ethics* 13 (4): 416–427. https://doi.org/10.1191/0969733006ne880oa.

Kälvemark Sporrong, S., A. Hoglund, M. Hansson, P. Westerholm, and B. Arnetz. 2004. Living with conflicts-ethical dilemmas and moral distress in the health care system. *Social Science and Medicine* 58 (6): 1075–1084. https://doi.org/10.1016/S0277-9536(03)00279-X.

Kälvemark Sporrong, S.K., A.T. Hoglund, M.G. Hansson, P. Westerholm, and B. Arnetz. 2005. "We are white coats whirling round"-Moral distress in Swedish pharmacies. *Pharmacy World & Science* 27 (3): 223–229. https://doi.org/10.1007/s11096-004-3703-0.

Karanikola, M.N.K., J.W. Albarran, E. Drigo, M. Giannakopoulou, M. Kalafati, M. Mpouzika, G.Z. Tsiaousis, and E.D.E. Papthanassoglou. 2013. Moral distress, autonomy and nurse physician collaboration among intensive care unit nurses in Italy. *Journal of Nursing Management* 22 (4): 472–484. https://doi.org/10.1111/jonm.12046.

Kelly, B. 1998. Preserving moral integrity: A follow-up study with new graduate nurses. *Journal of Advanced Nursing* 28: 1134–1145.

Khoiee, E.M., M. hossein Vaziri, S. Alizadegan, S.A. Motevallian, O.M. Razzaghi kashani, S. Ashrafoddin Goushegir, and J. Ghoroubi. 2008. Developing the moral distress scale in the population of Iranian nurses. *Iranian Journal of Psychiatry* 3 (2): 55–58.

Kilcoyne, M., and M. Dowling. 2008. Working in an overcrowded accident and emergency department: Nurses' narratives. *Australian Journal of Advanced Nursing* 25 (2): 21–27.

Ko, H.K., H.C. Tseng, C.C. Chin, and M.T. Hsu. 2019. Phronesis of nurses: A response to moral distress. *Nursing Ethics.* https://doi.org/10.1177/0969733019833126.

Laurs, L., A. Blaževičienė, E. Capezuti, and D. Milonas. 2019. Moral distress and intention to leave the profession: Lithuanian nurses in municipal hospitals: Original manuscript. *Journal of Nursing Scholarship:* 1–9. https://doi.org/10.1111/jnu.12536

Lazzari, T., S. Terzoni, A. Destrebecq, L. Meani, L. Bonetti, and P. Ferrara. 2019. Moral distress in correctional nurses: A national survey. *Nursing Ethics* 27 (1): 40–52. https://doi.org/10.1177/0969733019834976.

LeBaron, V., S.L. Beck, F. Black, and G. Palat. 2014. Nurse moral distress and cancer pain management: An ethnography of oncology nurses in India. *Cancer Nursing* 37 (5): 331–344. https://doi.org/10.1097/NCC.0000000000000136.

Lee, K.J., and C.Y. Dupree. 2008. Staff experiences with end-of-life care in the pediatric intensive care unit. *Journal of Palliative Medicine* 11 (7): 986–990. https://doi.org/10.1089/jpm.2007.0283.

Lomis, K.D., R.O. Carpenter, and B.M. Miller. 2009. Moral distress in the third year of medical school: A descriptive review of student case reflections. *American Journal of Surgery* 197 (1): 107–112. https://doi.org/10.1016/j.amjsurg.2008.07.048.

Lützén, K., A. Johansson, and G. Nordstrom. 2000. Moral sensitivity: Some differences between nurses and physicians. *Nursing Ethics* 7 (6): 520–530.

Maluwa, V.M., J. Andre, P. Ndebele, and E. Chilemba. 2012. Moral distress in nursing practice in Malawi. *Nursing Ethics* 19 (2): 196–207. https://doi.org/10.1177/0969733011414968.

Mänttäri-van der Kuip, M. 2019. Conceptualising work-related moral suffering—Exploring and refining the concept of moral distress in the context of social work. *The British Journal of Social Work:*1–17 https://doi.org/10.1093/bjsw/bcz034.

McCarthy, J. 2003. Principlism or narrative ethics: Must we choose between them? *Medical Humanities* 29:65–71. https://doi.org/10.1136/mh.29.2.65.

McCarthy, J., and C. Gastmans. 2015. Moral distress: A review of the argument-based nursing ethics literature. *Nursing Ethics* 22 (1): 131–152. https://doi.org/10.1177/09697330145571.

Mitton, C., S. Peacock, J. Storch, N. Smith, and E. Cornelissen. 2010. Moral distress among health-care managers: Conditions, consequences, and potential responses. *Healthcare Policy: Health Services, Management and Policy Research* 6 (2): 99–116.

Mueller, P.S., A.L. Ottenberg, D.L. Hayes, and B.A. Koenig. 2011. "I felt like the angel of death": Role conflicts and moral distress among allied professionals employed by the US cardiovascular implantable electronic device industry. *Journal of Interventional Cardiac Electrophysiology* 32: 253–261. https://doi.org/10.1007/s10840-011-9607-8.

Mukherjee, D., R. Brashler, T.A. Savage, and K.L. Kirschner. 2009. Moral distress in rehabilitation professionals: Results from a hospital ethics survey. *PM&R (Official journal of the American Academy of Physical Medicine and Rehabilitation)* 1 (5): 450–458.

Nathaniel, A.K. 2006. Moral reckoning in nursing. *Western Journal of Nursing Research* 28 (4): 419–438. https://doi.org/10.1177/0193945905284727.

Neumann, J.L., L.W. Mau, S. Virani, E.M. Denzen, D.A. Boyle, N.J. Boyle, J. Dabney, A. DeKesel Lofthus, M. Kalbacker, T. Khan, N.S. Majhail, E.A. Murphy, P. Paplham, L. Parran, M.A. Perales, T.H. Rockwood, K. Schmit-Pokorny, T.D. Shanafelt, E. Stenstrup, W.A. Wood, and L.J. Burns. 2018. Burnout, moral distress, work–life balance, and career satisfaction among hematopoietic cell transplantation professionals. *Biology of Blood and Marrow Transplantation* 24 (4): 849–860. https://doi.org/10.1016/j.bbmt.2017.11.015.

Ohnishi, K., Y. Ohgushi, M. Nakano, H. Fujii, H. Tanaka, K. Kitaoka, J. Nakahara, and Y. Narita. 2010. Moral distress experienced by psychiatric nurses in Japan. *Nursing Ethics* 17 (6): 726–740. https://doi.org/10.1177/0969733010379178.

Oliver, D. 2018. David Oliver: Moral distress in hospital doctors. *BMJ: British Medical Journal (Online)* 360. https://doi.org/10.1136/bmj.k1333.

Pauly, B.M., C. Varcoe, and J. Storch. 2012. Framing the issues: Moral distress in health care. *HEC Forum* 24 (1): 1–11. https://doi.org/10.1007/s10730-012-9176-y.

Pauly, B., C. Varcoe, J. Storch, and L. Newton. 2009. Registered nurses' perceptions of moral distress and ethical climate. *Nursing Ethics* 16 (5): 561–573. https://doi.org/10.1177/096973300 9106649.

Pendry, P.S. 2007. Moral distress: Recognizing it to retain nurses. *Nursing Economics* 25 (4): 217–221.

Penny, N.H., T.L. Ewing, R.C. Hamid, K.A. Shutt, and A.S. Walter. 2014. An investigation of moral distress experienced by occupational therapists. *Occupational Therapy in Health Care* 28 (4): 382–393. https://doi.org/10.3109/07380577.2014.933380.

Peter, E., and J. Liaschenko. 2004. Perils of proximity. A spatiotemporal analysis of moral distress and moral ambiguity. *Nursing Inquiry* 11 (4): 218–225. https://doi.org/10.1111/j.1440-1800. 2004.00236.x.

Pijl-Zieber, E.M., O. Awosoga, S. Spenceley, B. Hagen, B. Hall, and J. Lapins. 2018. Caring in the wake of the rising tide: Moral distress in residential nursing care of people living with dementia. *Dementia* 17 (3): 315–336. https://doi.org/10.1177/1471301216645214.

Pike, A.W. 1997. Moral outrage and moral discourse in nurse-physician collaboration. *Journal of Professional Nursing* 7 (6): 351–362.

Ramber, B., C. Vallett, J.A. Cohen, and J. Tarule. 2010. The moral cascade: Distress, eustress, and the virtuous organization. *Journal of Organizational Moral Psychology* 1 (1): 41–54.

Rawas, H. 2019. Moral distress in critical care nurses: A qualitative study. *International Journal of Studies in Nursing* 4 (4): 35–41. https://doi.org/10.20849/ijsn.v4i4.659.

Repenshek, M. 2009. Moral distress: Inability to act or discomfort with moral subjectivity. *Nursing Ethics* 16 (6): 734–742. https://doi.org/10.1177/0969733009342138.

Rodney, P. 1997. *Toward connectedness and trust: Nurses' enactment of their moral agency within an organizational context*. Vancouver, BC: University of British Columbia Editor.

Rushton, C. 1992. Care-giver suffering in critical care nursing. *Heart and Lung* 21: 303–306.

Rushton, C.H. 2013. Principled moral outrage: An antidote to moral distress? *American Association of Critical Care Nurses Advanced Critical Care* 24 (1): 82–89.

Sasso, L., A. Bagnasco, M. Bianchi, V. Bressan, and F. Carnevale. 2016. Moral distress in undergraduate nursing students: A systematic review. *Nursing Ethics* 23 (5): 523–534. https://doi.org/10.1177/0969733015574926.

Schoot, T., I. Proot, M. Legius, R. ter Meulen, and L. de Witte. 2006. Client-centred home care: Balancing between competing responsibilities. *Clinical Nursing Research* 15 (4): 231–254. https://doi.org/10.1177/1054773806291845.

Schwenzer, K.J., and L. Wang. 2006. Assessing moral distress in respiratory care practitioners. *Critical Care Medicine* 34 (12): 2967–2973. https://doi.org/10.1097/01.CCM.0000248879.190 54.73.

Severinsson, E. 2003. Moral stress and burnout: Qualitative content analysis. *Nursing & Health Sciences* 5 (1): 59–66.

Shoorideh, F.A., T. Ashktorab, and F. Yaghmaei. 2012. Response of ICU nurses to moral distress: A qualitative study. *Iranian Journal of Critical Care Nursing* 4 (4): 159–168.

Silén, M., M. Svantesson, S. Kjellström, B. Sidenvall, and L. Christensson. 2011. Moral distress and ethical climate in a Swedish nursing context: Perceptions and instrument usability. *Journal of Clinical Nursing* 20 (23–24): 3483–3493. https://doi.org/10.1111/j.1365-2702.2011.03753.x.

Somerville, M. 2000. *The ethical canary: Science, society and the human spirit*. New York, NY: Penguin Press.

Sundin-Huard, D., and K. Fahy. 1999. Moral distress, advocacy and burnout: Theorizing the relationships. *International Journal of Nursing Practice* 5: 8–13.

Ulrich, C.M., A.B. Hamric, and C. Grady. 2010. Moral distress: A growing problem in health professions? *Hastings Centre Report* 40 (1): 20–22. https://doi.org/10.1353/hcr.0.0222.

Varcoe, C., B. Pauly, G. Webster, and J. Storch. 2012. Moral distress: Tensions as springboards for action. *HCE Forum* 24 (1): 51–62. https://doi.org/10.1007/s10730-012-9180-2.

Vargas, I., and C. Concha. 2019. Moral distress, sign of ethical issues in the practice of oncology nursing: Literature review. *Aquichan* 19 (1): 11–15. https://doi.org/10.5294/aqui.2019.19.1.3.

Walker, M.U. 1993. Keeping moral space open: New images of ethics consulting. *The Hastings Center Report* 23 (2): 33–40.

Walsh, A. 2010. Pulling the heartstrings, arguing the case: A narrative response to the issue of moral agency in moral distress. *Journal of Medical Ethics* 36 (12): 746–749. https://doi.org/10.1136/jme.2010.036079.

Webster, G.C., and F. Baylis. 2000. Moral residue. In *Margin of error: The ethics of mistakes in the practice of medicine*, ed. S.B. Rubin and L. Zoloth, 217–230. Haggerstown: University Publishing Group.

Weinberg, M. 2009. Moral distress: A missing but relevant concept for ethics in social work. *Canadian Social Work Review/Revue Canadienne de Service Social* 26 (2): 139–152.

White, N., N. Beardslee, D. Peters, and J. Supples. 1990. Promoting critical thinking skills. *Nurse Educator* 15 (5): 6–19.

Whitehead, P.B., E.G. Epstein, and A.B. Hamric. 2018. Updating and refining a measure for moral distress: Introducing the MDS-2017. *Nursing Education Research Conference 2018 (NERC18)*. Retrieved from https://sigma.nursingrepository.org/bitstream/handle/10755/623871/Whitehead_Info_89562.pdf?sequence=2&isAllowed=y.

Wiggleton, C., E. Petrusa, K. Loomis, J. Tarpley, M. Tarpley, M.L. O'Gorman, and B. Miller. 2010. Medical students' experiences of moral distress: Development of a web-based survey. *Academic Medicine* 85 (1): 111–117.

Wilkinson, J. 1987. Moral distress in nursing practice: Experience and effect. *Nursing Forum* 23 (1): 16–29.

Wilson, M.A., D.M. Goettemoeller, N.A. Bevan, and J.M. McCord. 2013. Moral distress: Levels, coping and preferred interventions in critical care and transitional care nurses. *Journal of Clinical Nursing* 22 (9–10): 1455–1466. https://doi.org/10.1111/jocn.12128.

Wocial, L.D., and M.T. Weaver. 2012. Development and psychometric testing of a new tool for detecting moral distress: The moral distress thermometer. *Journal of Advanced Nursing* 69 (1): 167–174. https://doi.org/10.1111/j.1365-2648.2012.06036.x.

Wojtowicz, B., B. Hagen, and C. Van Daalen-Smith. 2014. No place to turn: Nursing students' experiences of moral distress in mental health settings. *International Journal of Mental Health Nursing* 23 (3): 257–264. https://doi.org/10.1111/inm.12043.

Zabetian, H., M.J. Zarei, F.H. Jahromy, S. Siasi, M. Radmehr, and S. Abiri. 2019. Investigation of moral distress in nurses of Jahrom hospitals in 2018. *Journal of Research in Medical and Dental Science* 7 (1): 195–200.

Zheng, R.S., Q.H. Guo, F.Q. Dong, and R.G. Owens. 2015. Chinese oncology nurses' experience on caring for dying patients who are on their final days: A qualitative study. *International Journal of Nursing Studies* 52 (1): 288–296. https://doi.org/10.1016/j.ijnurstu.2014.09.009.

Zuzelo, P. 2007. Exploring the moral distress of registered nurses. *Nursing Ethics* 14: 344–359. https://doi.org/10.1177/0969733007075870.

Chapter 2
Understanding Complex Human Responses

Abstract Moral uncertainty should be a reasonable expectation for contemporary health care providers who experience complex ethical issues in their everyday practice. The busy nature of caring work, technological and scientific advancements, and a multitude of competing demands leave little room for deep reflection on the complex moral aspects of patient and family care. From a western perspective, individualism and autonomy seem to be the dominant ethical values. It may be perceived that contemporary western society lives in an 'anything goes' type of culture where solid lines between right and wrong, good and bad, can sometimes blur. Moral disorientation, compassion fatigue, and burnout can and do occur. Also discussed here will be paradoxical concepts of moral discernment, compassion satisfaction, and moral courage. Each of these complex human responses represent an opportunity for renewed purpose and meaning in the work of caring for others.

Keywords Moral uncertainty · Moral disorientation · Compassion fatigue ·
Burnout · Compassion satisfaction

2.1 Introduction

The notion that nurses know what to do in the midst of an ethical quandary is a longstanding criticism of Jameton's (1984) original concept of moral distress. Is it possible for a nurse (or other health care provider) to experience moral distress without having a sense of what the right approach should be? To further explore these ideas, there is a need to go beyond solely cognitive components and include emotional and perhaps even intuitive aspects of ethical issues and ethical decision-making. Not fully knowing or understanding an ethical situation combined with apprehension in trusting one's own instinctive reactions sets the stage for moral uncertainty. This can occur in the moment or when looking back on an ethical situation, sometimes even years later, and asking:

What should I have done?

What else could I have done?

© Springer Nature Switzerland AG 2020

K. Jones-Bonofiglio, *Health Care Ethics through the Lens of Moral Distress*,
The International Library of Bioethics 82,
https://doi.org/10.1007/978-3-030-56156-7_2

Did I do the right thing after all?

Perhaps these introspective questions sound familiar. Would you ask yourself these questions if you knew the outcome of an ethical decision that you had been part of? If an outcome is deemed to be good, you might feel justified with the actions and decision making process that took place. Perhaps you would feel a warm glow of satisfaction that you made a positive contribution. Yes, you did this! However, it is important to be cautious.

If you were to find out that the final result was an unintended negative outcome you may feel guilty, angry, or even sad upon reflection. Perhaps, another kind of reaction comes to your mind. Yes, it was you who did this. Here it is also important to be cautious. This type of critical evaluation can be overly congratulatory or overly harsh as decision-making is being judged after-the-fact; a hindsight is 20/20 effect. Ruminations about '*if only I did/did not do this…*' may occur where alternate circumstances are explored for their hypothetical potential to mitigate a poor outcome that has already happened. This process is referred to informally as health care providers 'shoulding' all over themselves, in that they regretfully review what they should have done instead. Some health care providers may even habitually reflect back on the details of a particular ethical issue for decades after it occurs. The experience, in essence, haunts them. Missing here is a reality check that, for the most part, we make decisions and take action based on the limited information that we have at the time with (generally speaking) good intentions. We do the best we can with what we've got and most of the time we should not take all of the credit or all of the blame.

Other times, final outcomes are not known and these situations are often the most difficult to cope with. A health care provider may be left wondering about the final outcome and if their decisions possibly contributed to avoidable suffering. Facing unknown outcomes can lead to psychological rumination (Vandevala et al. 2017) about our direct and indirect moral contribution to others' lives. Reflection about the unknown can challenge our sense of confidence and competence, leading to experiences of moral uncertainty.

In my own graduate research on moral distress among Canadian nurses, it was nurses with decades of experience and wisdom that often expressed moral uncertainty, related to their most harrowing ethical issues (Jones-Bonofiglio 2015). Clearly, this was not due to a lack of practical knowledge and nursing expertise. Reminiscing about patients that they said they could not forget, they recounted situations as if they happened yesterday. Their collective stories revealed a deep sadness expressed with various sentiments that translated into, "*I still don't know if that was the right thing to do.*"

Ethics is a term that has a broad umbrella and it can include formal processes to address a moral dilemma or simply offering respect and dignity to a client and family. Ethics is (almost) always held in high regard, but it may feel intangible at times. Perhaps it is just assumed to be there, despite little thought or concerted effort. "*Of course I am practicing ethically.*" It is no wonder that ethics in practice may not always be top of mind at all times for busy health care providers. This lack of conscious attention or awareness may be for very good reasons. Nurses' everyday workplace stressors often include heavy workloads, unsupportive leadership, conflicts with coworkers, shift work, injuries (sometimes assaults), as just a few examples (Ross et al. 2018). These factors are similar to research findings on workplace stressors among emergency first responders (i.e., police, fire, paramedic); the stressors are both operational and organizational (Brough 2004). However for nurses, personal regrets about times when they felt that they were not able to be (fully) ethical as a nurse create circumstances where operational and organizational stressors have added and significant moral impact. Thus, there are many reasons why important moral issues are left unattended on the back burner to simmer and, eventually, to smolder.

2.2 Moral Uncertainty and Disorientation

Moral uncertainty can be generally described as a state of being unsure of how to act or proceed with an ethical issue. Often conceptualized as a multidimensional experience, it can be accompanied with feelings of isolation, shock, overwhelm, or surprise (Harbin 2012). Distilled, it is an activation of one's scruples. However, Jameton (1984) restricted the use of the term moral uncertainty to circumstances with an unclear moral action or unclear ethical issues, and separated it from experiences of moral distress. Here I argue that the second-guessing that occurs in moral uncertainty can lead to an erosion of one's sense of moral confidence and the resulting hesitation to act may also be a source of moral distress. Thus there may be an important connection between moral distress and moral uncertainty. This does not ring true with Jameton's original work as he clearly defined the two concepts separately. However, a potential interrelationship is being put forward here as a consideration for broadening our understanding of moral distress and has been suggested by others.

For example, a recent concept analysis of moral distress and moral uncertainty in nursing found these two concepts to be distinct from each other, however described an interconnection (Dorman and Raffin Bouchal 2020). Moral distress had the standard definition, but unique attributes of moral uncertainty were noted to include internal conflict related to circumstances with ambiguous outcomes. In moral conflict, nurses struggled with questions about the 'right' course of action and with articulating the conflict itself. The end result was often silence and inaction. Outcomes of moral uncertainty included experiencing a loss of control and negative emotions (e.g., frustration, anger, depression, anxiety). At best this may result in a laissez-faire attitude that simply supports the status quo through dependence on an external locus of control. At worst, nurses feel powerless and remorseful for perceived wrongdoing, internalizing the effects of the experience. Ongoing experiences of moral uncertainty and acceptance of these uncomfortable circumstances may lead to indifference. These Canadian authors consider moral uncertainty and moral distress to be unique concepts with an iterative, spiralling relationship.

Recap of Concept: Moral Uncertainty
- ✓ Moral incongruence about an ethical issue
- ✓ No single correct option for ethical action
- ✓ Uncertainty about how to proceed
- ✓ Internal conflict
- ✓ A sense of loss of control
- ✓ Inaction and an experience of negative emotions.

Moral disorientation is, perhaps, a less common term, but it describes a common human experience of feeling lost (Reid 2014). It occurs on a spectrum of disconnection. A loss of grounding on a firm moral foundation can lead to a variety of responses from stringent adherence to the rules to complete disengagement in an effort to regain a lost sense of control. Experiences of ongoing moral distress, moral uncertainty, and moral disorientation can be corrosive to an individual's sense of self and self-efficacy if left unaddressed.

But what about middle ground experiences on the spectrum of moral disorientation? Sometimes there is room during or after a disorienting experience to shift one's attention and allow new perspectives and/or opportunities for reflecting on moral experiences in a more (or less) intense way. Disorientations can help us to understand ourselves and others better, as experiences of vulnerability can change how we pay attention to the moment at hand (Harbin 2012).

Canadian nursing scholar Graham McCaffrey (2019) highlights nursing as a relational discipline that is specific and focused on the suffering of the other in complex environments. Due to this relationship, he writes that health care professionals operate on an 'ethical tilt' and must assume ethical responsibilities that are unbalanced with those for whom they care. The moral weight of the work of being a care provider allows for a unique and vulnerable position to (potentially) see an ethical issue from a kaleidoscope of perspectives, as providers are present, bear empathetic witness, and attend to fellow human beings' needs (Harbin 2012). Moral distress experiences can present a 'liminal space' or 'threshold moment', where health care providers feel they cannot go back from and are unsure of how to take the next steps forward.

Further, there seems to be a keen sense of disorientation occurring in the debate about how to define the concept of moral distress itself. In fact some researchers, such as Megan-Jane Johnstone (Australia's leading nursing ethics scholar), believe it may be time to toss out the concept of moral distress entirely. Johnstone and her colleague, Alison Hutchinson, highlight the theoretical and practical flaws inherent in the moral distress research and suggest that a focus on better moral decision making would be a far more productive endeavour (Johnstone and Hutchinson 2015).

2.3 Moral Discernment

Morals are the general principles of right and wrong, of good behaviour, and of proper character. Moral decisions may require the application of these principles to a particular situation. Moral decision-making involves using moral principles and ethical reasoning to rationalize and defend one's choice of action or non-action. It is important to recognize that there are many factors to be considered by the decision maker, including context, history, relationships, character, and culture (Thompson 2002).

What if, in the application of principles to a moral decision, there is more than one option and the right course of action is unclear? This is the tricky circumstance of moral conflict. A moral conflict may also present within an ethical dilemma where one must choose between two equally undesirable alternatives (Jameton 1984). A moral dilemma can be defined as a circumstance where no single right answer is available (Begley 2008). This type of moral maze represents the proverbial scenario of being between 'a rock and a hard place'. Such experiences can represent crucible moments, where one feels it could 'make or break' them. A search for a suitable compromise may ensue. It may be a simple compromise of interest or it may require a moral compromise that impacts integrity and/or identity. Further, it is often difficult to fairly weigh the value of the requirements of others' interests against one's own self interest.

According to the Oxford online dictionary (2019), 'discernment' is the ability to judge a situation well. Moral discernment involves being primed by sensitivity to the ethical dimensions of self, others, and context (Rushton et al. 2013). It involves more that just knowing the facts or the step-by-step application of principles in an algorithmic-like fashion. It is complementary to rational and objective decision making. Moral discernment can include following the rules or, sometimes, the need to break the rules (for example, in the name of social justice and equity). It can occur when we do the right thing for the right reasons and when we do the 'wrong' thing for the right reasons.

Discernment also includes recognizing and responding to our own emotions and emotional reactions to a situation. Our emotions are noted to be highly valuable guides to moral action as they are principled judgments with insights into what we think really matters (Keltner 2009). The things that matter, that we have learned to care about most, is our cathexis (Seligman 2011). Our cathexis is where we choose to place our time and energy and, hopefully, where we find meaning and purpose.

In nursing there is such a concept as a 'good nurse'. In an American study the meaning of being a good nurse was explored among 20 registered nurses across three hospitals (Catlett and Lovan 2011). Four key themes emerged from the data, including: (1) caring/caring behaviours; (2) personal attributes; (3) skills and knowledge; and, (4) work environment/coworkers. There was some overlap between the four categories and all aspects spoke to the need for a 'good nurse' to get the job done well, efficiently, effectively, and ethically. These researchers recommend that schools of nursing and nurse employers looking to hire new nurses should pay particular attention to these valuable, and perhaps essential, qualities of a 'good nurse'.

Further, moral discernment is a skill that can be honed over time and with experience. Discernment takes ethical decision-making one step further and includes listening to one's sense of inner knowing of what feels instinctually good and right. Some may equate the language of discernment with Christian religious traditions and thus discernment may not be considered an accessible concept for all. However, discernment in the context of this book does not necessitate a belief in the divine guidance of a god or creator unless the reader so chooses.

2.4 Common Moral Commitments

Nurses are expected to practice ethically for the good of their patients. In a recent concept analysis of moral courage in nursing, the antecedents for 'taking the tough stand' of moral courage were found to include experience, conscience, and ethical sensitivity (Numminen et al. 2017). Therefore, moral courage is about more than just a sense of bravery. It offers the potential for empowerment and personal growth. So why is it that in the face of moral injustice, health care providers often fail to act? What factors might support moral courage in practice (Gallagher 2010)?

While theories and approaches to ethics and moral values are often separated into particular genres, there are common moral commitments that various disciplines can generally agree on. Further, there is an argument that congruent, and perhaps even universal, moral beliefs exist and are held within and across all health care professions. It is not surprising to find such common ground given that the collective work of health care is meant to benefit others (e.g., patients, clients, residents, persons receiving care, etc.).

Canadian ethics scholar pioneers, Janet Storch (nursing) and Nuala Kenny (medicine) write about the shared moral work of nurses and physicians in the context of recognizing the value of collaboration and the strengths of unique sets of knowledge, skills, and ethical perspectives (Storch and Kenny 2007). Historically, nurses had early moral commitments which were focused largely on etiquette and loyalty to their primary role as hand-maiden of the physician, with holistic commitments to patients' body, mind, and soul. Physicians have seen their moral work shift over time from a priority need to present as a 'gentleman' toward an overarching focus on clinical skills and scientifically-based judgment in order to meet the needs of their patients. Storch and Kenny (2007) call for the internalization of ethics in practice (beyond just codes of ethics) and respectful collaboration toward a shared moral enterprise of duties and obligations among physicians and nurses.

In a very large Swedish study, the moral sensitivity to ethical issues in practice of 1073 physicians and nurses was explored using the Moral Sensitivity Questionnaire (MSQ) (Lützén et al. 2000). Researchers found significant differences in the responses of the two groups in the category of 'benevolence'; understood as kindness or good will. They found that nurses and physicians had divergent perspectives on acts considered to be kind or for the good of the other. One item in particular stood out as the most extreme difference of perspective and that was on the topic of the threat of forced medication administration or coercion. Nurses did not support this as a benevolent act. Further, there was a gender difference as well with males being more likely to have a positive attitude toward coercion. However, other areas explored by the MSQ, such as relationships, meaning, autonomy, conflict, and rules revealed no statistically significant differences between the self-reported views of the physicians and nurses in this study.

2.5 Moral Outrage

If there can be a case made for common moral commitments, there is also potential for shared experiences of moral outrage among health care providers. Moral outrage can be defined as outwardly directed and justified anger in response to the belief that an ethical breach has occurred (Pike 1991; Rushton 2013). Moral outrage can begin with a new or an all-too-familiar sense of disorientation; a sense that something is not right here. This builds an emotionally intense response that can be useful for proceeding with ethical action, but not always.

Cynda Rushton, an American nurse scholar and ethicist, notes that there is an important difference between ungrounded moral outrage and principled moral outrage (Rushton 2013). Ungrounded moral outrage, is outwardly directed, loosely justified anger without acknowledgement of ethical principles or attempts to define the moral issue itself. It is an intuitive, gut-level reaction to real or perceived injustice. Principled (or grounded) moral outrage is also outwardly directed, justified anger but it takes a more thoughtful, reasoned approach to explore the contours of an ethical issue with wisdom and compassion. Thus moral outrage, offers an opportunity for moral discernment to be diminished or enhanced depending on an individual's approach.

The response of ungrounded moral outrage may restrict access to specific regions of the brain for decision-making and processing uncertainty. Negative emotions, such as anger, are most closely associated with the right prefrontal cortex, whereas positive emotions show access to the left prefrontal cortex (Halifax 2011). As anger festers and builds toward a state of rage, rational, non-dualistic thinking may be limited. Further, anger as an intense emotional response encourages the brain to hold onto the experience and this increases the risk for rumination through a focus on one's own sense of distress and injustice. This is problematic because it interferes with the primary role of the professional care provider to hold the patient through their suffering, not to hold the suffering for themselves (Kerney 2000). If suffering is routinely held, health care providers risk experiencing compassion fatigue, burnout, and related sequela.

I argue that moral outrage is often a short-term opportunity, as it is an energy draining experience (Pike 1991). This level of hyper-arousal, even when grounded in principles, requires fight and fuel. The fuel for the fight may be sourced internally or externally, and such resources are unique to each individual. As experiences of real or perceived injustice(s) build up, moral outrage suggests a focus on an external locus of control (because if one had the ability to fix/stop the injustices, they would not continue to occur). A vantage point to locate self in the mix can be missing. This restricts the possibilities that may be grasped for effective ethical decision-making and does not offer a full view orientation.

2.6   Compassion Fatigue

The concept of compassion fatigue among professional caregivers was first described in the academic literature by registered nurse Carla Joinson almost 30 years ago. She noted it among emergency room nurses as a syndrome of burnout that resulted in a lack of ability to nurture (Joinson 1992). Further, American psychologist Charles Figley (1995) is well cited for defining compassion fatigue (among psychotherapists and counsellors) as being equivalent to secondary traumatic stress disorder and later noted it as a form of post traumatic stress disorder (PTSD) (Figley 2002).

Compassion fatigue is generally described as the cumulative result of responding to suffering with empathy and engagement. It is thought to occur over a prolonged period of time (repeated exposure), however it is now recognized even after a one time event (American Institute of Stress 2019). Compassion fatigue has been identified among individuals and also as a more collective cultural experience among a community in a particular environment (Austin et al. 2009). It is the resulting trauma of ongoing and authentic personal engagement that eventually evolves into, both, hopelessness and helplessness. Symptoms of compassion fatigue can fall into three general categories which include: (1) negative responses to work; (2) reduced empathy; and, (3) lack of joy (Lombardo and Eyre 2011). Further, there are many other terms in the academic literature that are similar to or perhaps interchangeable with compassion fatigue (see Table 2.1).

Shane Sinclair, a Canadian professor with a background in theology, is internationally recognized as a leading expert on compassion in health care. In a meta-narrative review of the health care literature, Sinclair and his colleagues describe compassion fatigue as an acute onset stress response due to occupational factors with both physical and emotional manifestations (Sinclair et al. 2017). Compassion fatigue associated with vicariously experiencing the suffering of others has been identified among many types of professionals (see Table 2.2)

Canadian nursing ethics scholar and former Canada Research Chair in Relational Ethics in Health Care (2003–2013), Wendy Austin et al. (2009) write about compassion fatigue among nurses as extending beyond their professional roles, into their personal lives, and challenging both their identity and sense of hope. They highlight that compassion fatigue is thought to have individual provider antecedents (i.e., psychological make-up, nature of work, how engagement is managed) and also be related to environmental conditions (i.e., globalization, media, market rationalization). In their research, interviews with five nurses revealed six interconnected

Table 2.1 Terms related to compassion fatigue[a]

Terms	Examples from the academic literature
Burnout	Burnham et al. (2019), Collins and Long (2003), Dzeng and Curtis (2018), Edwards et al. (2000), Pines (1993), Pines and Aronson (1988), Severinsson (2003)
Countertransference	Fox and Carey (1999)
Empathic distress	Wacker and Dziobek (2018)
Empathic distress fatigue	Klimecki and Singer (2012)
Moral stress	Lützén et al. (2003, 2006, 2000)
Occupational stress	Kakiashvili et al. (2013), Santos e al. (2010)
Post traumatic stress disorder (PTSD)	Czaja et al. (2012), Figley (2005)
Secondary trauma	Coholic and Blackford (2005)
Secondary traumatic stress (STS)	Kelly et al. (2015), Meadors et al. (2010), Munroe et al. (1995), Salston and Figley (2003)
Traumatic countertransference	Herman (1992)
Vicarious traumatization	McCann and Pearlman (1990)

[a]Additional terms include: compassion stress, complex emotional burdens (CEBs), emotional contagion, empathy fatigue, multiple psychic wounds, operational stress injury (OSI), post-traumatic stress-like symptoms, primary traumatic stress, secondary traumatization, and secondary victimization (co-victimization)

themes for compassion fatigue that involved feeling emotionally empty, withdrawing, feeling impotent, losing balance, grieving the nurse they used to be, and just trying to survive (Austin et al. 2009). Further, participants described deliberate attempts to protect themselves from further exposure to the suffering of others with avoidance behaviours, due to feeling a sense of emptiness and hopelessness.

2.7 Burnout

As a closely related concept, burnout manifests with a sense of dissatisfaction with work. It is different from compassion fatigue, but the two can co-exist (American Institute of Stress 2019). First introduced as a concept in the late 1970s (Pines and Maslach 1978), burnout is defined as being multilayered and involves: (1) belief in a low level of personal accomplishment; (2) depersonalization behaviours; and, (3) feelings of emotional exhaustion.

The concept of burnout is characterized by a constellation of responses in an individual, which may include but are not limited to cynicism, decreased productivity, and exhaustion (Talbot and Dean 2018). According to the American Institute of Stress (2019), burnout involves four stages: (1) enthusiasm; (2) stagnation; (3)

Table 2.2 Exposure to the suffering of others[a] among various professions

Professions	Examples from the academic literature
Child welfare workers	Nelson-Gardell and Harris (2003), Salloum et al. (2015)
Disaster responders	Gustavsson et al. (2020)
Firefighters	Bastug et al. (2019)
First responders	Greinacher et al. (2019)
Lawyers	Leclerc et al. (2019), Maguire and Byrne (2017), Zwisohn et al. (2018)
Nurses	Kelly et al. (2015)
Nurses and community service workers	Cocker and Joss (2016)
Paramedics	Al Enazi and AlEnzie (2018)
Physicians	Huggard (2003)
Physiotherapists	Santos et al. (2010)
Police officers	Blumberg and Papazoglou (2019), Grant et al. (2018), Papazoglou et al. (2019)
Sexual assault workers	Coholic and Blackford (2005)
Social workers	Ashley-Binge and Cousins (2019), Wacker and Dziobek (2018)
Trauma therapists	Sexton (1999)
Teachers	Koenig et al. (2018)
Television journalists	Dworznik (2018)
Victim advocates	Cummings et al. (2018)

[a]Also referred to as burnout, secondary trauma, moral distress, post traumatic stress symptoms, compassion fatigue, occupational stress, empathic distress, secondary stress

frustration; and, (4) apathy. As such, burnout is linked to higher risks for poor physical health, namely cardiovascular and musculoskeletal diseases (Honkonen et al. 2006; Melamed et al. 2006) and poor mental health (Oskrochi et al. 2016).

While the terms burnout and compassion fatigue are often used interchangeably, burnout is uniquely considered to be a symptom of the brokenness of our contemporary health care system and a human response to unyielding organizational stressors and systemic pressures. Although the two concepts do share many symptoms, burnout apparently does not require trauma exposure or caring. Common symptoms to both burnout and compassion fatigue are a sense of emotional, mental, and/or physical exhaustion, finding reduced meaning in one's work, isolating self, and feeling disconnected from others (American Institute of Stress 2019). Protective factors for health care providers include having a higher education, psychological resilience, a moderate to high level of mindfulness, and good mental health; see Table 2.3 for individual-based factors and their proposed impacts.

Table 2.3 Factors related to compassion fatigue and burnout

Individual-based factors	Compassion fatigue	Burnout	PTSD symptoms	Secondary traumatic stress	Impact	References
Higher education	x	x			Reduced rates	Zhang et al. (2018)
Psychological resilience		x	x	x	Reduced prevalence	Mealer et al. (2012)
Low level of mindfulness		x			Strongest predictor	Viladarga et al. (2011)
Neuroticism		x			Strongly related	Rees et al. (2015)

2.8 Compassion Satisfaction

But, what of those health care providers who have developed a keen sense of moral awareness and various skills for moral discernment? These may include health care providers who have healed from experiences of compassion fatigue and/or burnout; those who have learned to thrive and not just to survive in complex and challenging working environments. Here we enter into the domain of compassion satisfaction; an altruistic quality of feeling self-appreciation from helping behaviours directed toward others (Zhang et al. 2018).

In terms of an individual health care provider's overall satisfaction, a key predictor of this is freedom of choice (Buettner 2010). For example, a Brazilian study of 141 nurses explored the connections between moral distress and work satisfaction (Wachholz et al. 2019). The most important component of satisfaction was found to be autonomy. Also important were factors such as interaction and remuneration. In term of moral distress, related components were autonomy, working conditions, and competence of team members.

Satisfaction is important in terms of ethical decision-making. If health care providers feel that they have very little input into or control over the moral decisions that are important to them, they will not be satisfied. Therefore compassion satisfaction is about the journey and not so much about the destination. In a meta-analysis of 11 studies about nurses, compassion satisfaction was strongly correlated with sociality and positive affect (Zhang et al. 2018). The quality of the journey itself depends on how adversity is processed and who is there to support you along the way. Similarly, a person's quality of life is closely tied by two key factors; experiences of accomplishment at work and relationships with others (Csikszentmihalyi 1990).

In an American study of almost 500 direct care nurses, predictors of compassion satisfaction included job satisfaction, meaningful recognition, and nurses who were either in late or early career (e.g., Baby Boomer or Millennial generations) (Kelly et al. 2015). Of note, late or early career nurses are generally the cohorts most represented in studies about moral distress and ethics in practice. Therefore, one has to wonder about the lack of nurses in mid career in studies about ethical issues, ethical decision making, moral distress, and now, compassion satisfaction. They hardly seem to be represented and they may have something incredibly valuable to share to further our understanding about these concepts.

2.9 Conclusion

In an effort to visualize the interrelationships between the concepts of moral courage, moral outrage, compassion fatigue, and burnout a model (Fig. 2.1) was created to posit possible connections and propose ways that levels of caring and moral distress might be influenced.

There is scholarly discontent with the traditionally accepted definitions of moral distress, compassion fatigue, and burnout. The arguments can be distilled to the notion that, perhaps, we are unfairly blaming individuals for completely natural responses to complex and difficult (sometimes inhumane), morally challenging issues. Therefore, responses to such circumstances should not be judged as a lack of ethical hardiness or moral toughness. As humans, should we not feel and experience the sorrow and suffering of others? In fact, science has shown that we are actually wired to do so (e.g., mirror neurons; Ferrari and Coudé 2018). Empathetic and compassionate responses are experiences of common humanity and solidarity that should be supported and nurtured.

It can be difficult to fully grasp an understanding of the complex human responses that individuals can have in response to the suffering of others. However, it would be misguided to exclusively seek satisfaction. Brene Brown (2017), an American research professor and social worker, writes about her mistake of trying to seek

Fig. 2.1 Responses of moral distress and caring

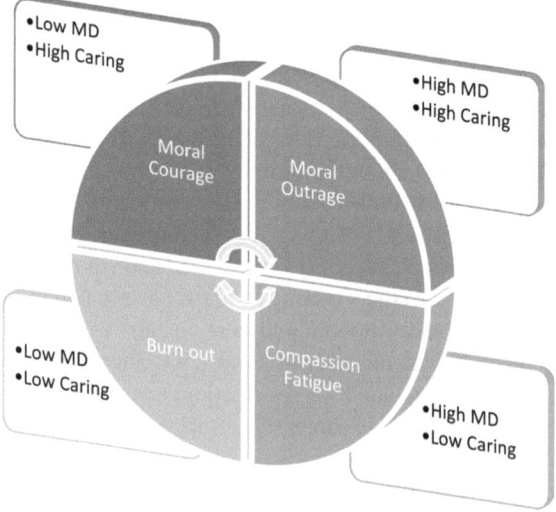

instant joy (e.g., fast, quick, easy) in her work instead of recognizing her pain, her authentic story, and her ability for compassion. Caring work can be intensely difficult and incredibly rewarding. Caring for others can also teach us about ourselves and our place in the world. Often though, the complexity of these experiences are overlooked and sometimes minimized.

Further, a dominant emphasis on individual-based interventions, such as self-care, mindfulness, and meditation, overlook the wider structural, organizational, and systemic dynamics that contribute to compassion fatigue, burnout, and moral distress experiences. As will be suggested throughout this book, meaningful and effective solutions will need to be multi-layered and range from micro to macro perspectives. However, I argue that it is also important to provide individuals with the knowledge and the power to make their own choices and changes in collaboration with efforts toward larger scale changes.

Other criticisms include the lack of attention to detail in terms of scientific proof that these concepts even actually exist (Sinclair et al. 2017). The use of multiple terms with variable definitions that often overlap with each other and share symptoms, muddies the water in the evidence-informed search for strategies to address (or at least mitigate) negative outcomes and is a barrier to solid evaluation attempts. Additionally, there may be assumptions that a desire to care for others is always a good thing. Pathological altruism is an interesting term used by researchers Oakley, Knafo, and McGrath (2011) to describe an action toward the wellbeing of another with unanticipated negative outcomes to the self and/or the other (otherwise described as 'helping that hurts').

The reality is that moral distress among nurses (and other health care providers, too) costs more than anyone can afford to lose. Compassion fatigue and burn out cost organizations a lot of money, even when employees still show up for work (presenteeism). Further, moral distress, compassion fatigue, and burnout contribute to ongoing issues with recruitment and retention of quality health care providers. These experiences can drive good health care providers out of workplaces (job turnover) or caring professions altogether (attrition). This burden is shared with patients and families in terms of not receiving the highest quality of care and experiencing the best possible health outcomes.

And, not least of which, unaddressed moral distress affects individual health care providers who hopefully arrived bright and shiny to their profession with goals to make a difference in the lives of others. How could they have anticipated that they would eventually feel so isolated, overwhelmed, and confused? Thus, presently there is a dire need for a moral community among nurses and other health care providers to support each other in quality ethical practice. There is a need for the collective strength of collaborative planning, communication, and interdependent work, in spite of society's privileging of individualism and independence (Brown 2017).

References

American Institute of Stress. 2019. *Compassion fatigue.* Retrieved from https://www.stress.org/mil itary/for-practitionersleaders/compassion-fatigue.

Al Enazi, S.K., and A.N. AlEnzie. 2018. Stress and burnout among Red Crescent paramedic ambulance workers in Riyadh. *Integrative Trauma and Emergency Medicine* 6: 2–10.

Ashley-Binge, S., and C. Cousins. 2019. Individual and organisational practices addressing social workers' experiences of vicarious trauma. *Practice* 1–17. https://doi.org/10.1080/09503153. 2019.1620201.

Austin, W., E. Goble, B. Leier, and P. Byrne. 2009. Compassion fatigue: The experience of nurses. *Ethics and Social Welfare* 3 (2): 195–214. https://doi.org/10.1080/17496530902951988.

Bastug, G., A. Ergul-Topcu, E. Tugba Ozel-Kizil, and O. Furkan Ergun. 2019. Secondary traumatization and related psychological outcomes in firefighters. *Journal of Loss and Trauma* 24 (2): 143–158. https://doi.org/10.1080/15325024.2018.1560898.

Begley, A.M. 2008. Truth-telling, honesty and compassion: A virtue-based exploration of a dilemma in practice. *International Journal of Nursing Practice* 14 (5): 336–341. https://doi.org/10.1111/ j.1440-172X.2008.00706.x.

Blumberg, D., and K. Papazoglou. 2019. A brief introduction to multiple psychic wounds in police work. *Crisis, Stress, & Human Resilience: An International Journal* 1 (1): 28–31.

Brough, P. 2004. Comparing the influence of traumatic and organizational stressors on the psychological health of police, fire, and ambulance officers. *International Journal of Stress Management* 11 (3): 227–244. https://doi.org/10.1037/1072-5245.11.3.227.

Brown, B. 2017. *Braving the wilderness: The quest for true belonging and the courage to stand alone.* New York, NY: Random House.

Buettner, D. 2010. *Thrive: Finding the happiness the blue zones way.* Washington, D.C.: National Geographic.

Burnham, E.L., K.E. Burns, M. Moss, and P.M. Dodek. 2019. Burnout in women intensivists: a hidden epidemic? *The Lancet Respiratory Medicine* 7 (4): 292–294. https://doi.org/10.1016/ S2213-2600(19)30029-3.

Catlett, S., and S.R. Lovan. 2011. Being a good nurse and doing the right thing: A replication study. *Nursing Ethics* 18 (1): 54–63. https://doi.org/10.1177/096973301038616.

Cocker, F., and N. Joss. 2016. Compassion fatigue among healthcare, emergency and community service workers: A systematic review. *International Journal of Environmental Research and Public Health* 13 (6): 618. https://doi.org/10.3390/ijerph13060618.

Coholic, D., and K. Blackford. 2005. Exploring secondary trauma in sexual assault workers in Northern Ontario locations. In *Violence in the family: Social work readings and research from northern and rural Canada*, eds. K. Brownlee and J.R. Graham; Toronto, ON: Canadian Scholar's Press.

Collins, S., and A. Long. 2003. Too tired to care? The psychological effects of working with trauma. *Journal of Psychiatric and Mental Health Nursing* 10: 17–27. https://doi.org/10.1046/ j1365-2850.2003.00526.x.

Csikszentmihalyi, M. 1990. *Flow: The psychology of optimal experience.* New York, NY: Harper Perennial.

Cummings, C., J. Singer, R. Hisaka, and L.T. Benuto. 2018. Compassion satisfaction to combat work-related burnout, vicarious trauma, and secondary traumatic stress. *Journal of Interpersonal Violence* 1–16. https://doi.org/10.1177/0886260518799502.

Czaja, A.S., M. Moss, and M. Mealer. 2012. Symptoms of posttraumatic stress disorder among pediatric acute care nurses. *Journal of Pediatric Nursing* 27 (4): 357–365. https://doi.org/10. 1016/j.pedn.2011.04.024.

Dorman, J.D., and S. Raffin Bouchal. 2020. Moral distress and moral uncertainty in medical assistance in dying: A simultaneous evolutionary concept analysis. *Nursing Forum* 1–11. https://doi. org/10.1111/nuf.12431.

Dworznik, G. 2018. Personal and organizational predictors of compassion fatigue symptoms in local television journalists. *Journalism Practice* 12 (5): 640–656. https://doi.org/10.1080/17512786. 2017.1338532.

Dzeng, E., and J.R. Curtis. 2018. Understanding ethical climate, moral distress, and burnout: a novel tool and a conceptual framework. *BMJ Qualitative Safety* 27: 766–770. https://doi.org/10.1136/ bmjqs-2018-007905.

Edwards, D., P. Burnard, D. Coyle, A. Fothergill, and B. Hannigan. 2000. Stress and burnout in community mental health nursing: a review of the literature. *Journal of Psychiatric and Mental Health Nursing* 7: 7–14. https://doi.org/10.1046/j.1365-2850.2000.00258.x.

Ferrari, P.F., and G. Coudé. 2018. Mirror neurons, embodied emotions, and empathy. In *Neuronal correlates of empathy: From rodent to human*, eds. K.Z. Meyza, and E. Knapska, 67–77. New York, NY: Academic Press. https://doi.org/10.1016/B978-0-12-805397-3.00006-1.

Figley, C.R. 1995. *Compassion fatigue: Coping with secondary traumatic stress disorder in those who treat the traumatized*. New York, NY: Bruner/Mazel.

Figley, C.R. 2002. *Coping with secondary traumatic stress disorder in those who treat the traumatized*. London, UK: Brunner-Routledge.

Figley, C.R. 2005. Strangers at home: Comment on Dirkzwager, Bramsen, Adèr, and van der Ploeg. *Journal of Family Psychology* 19 (2): 227–229. https://doi.org/10.1037/0893-3200.19.2.227.

Fox, R., and L.A. Carey. 1999. Therapists' collusion with the resistance of rape survivors. *Clinical Social Work Journal* 27 (2): 185–201.

Gallagher, A. 2010. Moral distress and moral courage in everyday nursing practice. *Online Journal of Issues in Nursing* 16 (2). https://doi.org/10.3912/OJIN.Vol16NO02PPT03.

Grant, H.B., C.F. Lavery, and J. Decarlo. 2018. An exploratory study of police officers: low compassion satisfaction and compassion fatigue. *Frontiers in Psychology* 9: 2793. https://doi.org/10. 3389/fpsyg.2018.02793.

Greinacher, A., C. Derezza-Greeven, W. Herzog, and C. Nikendei. 2019. Secondary traumatization in first responders: a systematic review. *European Journal of Psychotraumatology* 10 (1): 1562840. https://doi.org/10.1080/20008198.2018.1562840.

Gustavsson, M.E., F.K. Arnberg, N. Juth, and J. von Schreeb. 2020. Moral distress among disaster responders: what is it? *Prehospital and Disaster Medicine* 1–8. https://doi.org/10.1017/S10490 23X20000096.

Halifax, J. 2011. The precious necessity of compassion. *Journal of Pain and Symptom Management* 41 (1): 146–153. https://doi.org/10.1016/j.jpainsymman.2010.08.010.

Harbin, A. 2012. Bodily disorientation and moral change. *Hypatia* 27 (2): 261–280.

Herman, J. 1992. *Trauma and recovery*. New York, NY: Basic Books.

Honkonen, T., K. Ahola, M. Pertovaara, E. Isometsä, R. Kalimo, E. Nykyri, E., A. Aromaee, and J. Lönnqvist. 2006. The association between burnout and physical illness in the general population— Results from the Finnish Health 2000 Study. *Journal of Psychosomatic Research* 61 (1): 59–66. https://doi.org/10.1016/j.jpsychores.2005.10.002.

Huggard, P. 2003. Compassion fatigue: How much can I give? *Medical Education* 37: 163–164. https://doi.org/10.1046/j.1365-2923.2003.01414.x.

Jameton, A. 1984. *Nursing practice: The ethical issues*. Engelwood Cliffs, NJ: Prentice Hall.

Johnstone, M.J., and A. Hutchinson. 2015. 'Moral distress'– time to abandon a flawed nursing construct? *Nursing Ethics* 22 (1): 5–14. https://doi.org/10.1177/0969733013505312.

Joinson, C. 1992. Coping with compassion fatigue. *Nursing* 22(4):116,118–120.

Jones-Bonofiglio, K. 2015. *What guides us here? Exploring community health nurses' experiences of moral distress* (unpublished doctoral dissertation). Lakehead University, Thunder Bay, ON, Canada. Retrieved from https://knowledgecommons.lakeheadu.ca/handle/2453/672.

Kakiashvili, T., J. Leszek, and K. Rutkowski. 2013. The medical perspective on burnout. *International Journal of Occupational Medicine and Environmental Health* 26: 401–412. https://doi.org/ 10.2478/s13382-013-0093-3.

Kelly, L., J. Runge, and C. Spencer. 2015. Predictors of compassion fatigue and compassion satisfaction in acute care nurses. *Journal of Nursing Scholarship* 47 (6): 522–528. https://doi.org/10. 1111/jnu.12162.

Keltner, D. 2009. *Born to be good: The science of a meaningful life.* New York, NY: W.W. Norton Company.

Kerney, M. 2000. *A place of healing: Working with suffering in living and dying.* Oxford: Oxford University Press.

Klimecki O., and T. Singer. 2012. Empathic distress fatigue rather than compassion fatigue? Integrating findings from empathy research in psychology and social neuroscience. In *Pathological altruism,* eds. B. Oakley, A. Knafo, G. Madhavan, and D.S. Wilson, 369–383. New York, NY: Oxford University Press.

Koenig, A., S. Rodger, and J. Specht. 2018. Educator burnout and compassion fatigue: A pilot study. *Canadian Journal of School Psychology* 33 (4): 259–278.

Leclerc, M.E., J.A. Wemmers, and A. Brunet. 2019. The unseen cost of justice: post-traumatic stress symptoms in Canadian lawyers. *Psychology, Crime & Law* 1–21.

Lombardo, B., and C. Eyre. 2011. Compassion fatigue: A nurse's primer. *OJIN: The Online Journal of Issues in Nursing* 16 (1): Manuscript 3. https://doi.org/10.3912/OJIN.Vol16No01Man03.

Lützén, K., A. Cronqvist, A. Magnusson, and L. Andersson. 2003. Moral stress: synthesis of a concept. *Nursing Ethics* 10 (3): 312–322. https://doi.org/10.1191/0969733003ne608oa.

Lützén, K., V. Dahlqvist, S. Eriksson, and A. Norberg. 2006. Developing the concept of moral sensitivity in health care practice. *Nursing Ethics* 13 (2): 187–196. https://doi.org/10.1191/096 9733006ne837oa.

Lützén, K., A. Johansson, and G. Nordstrom. 2000. Moral sensitivity: Some differences between nurses and physicians. *Nursing Ethics* 7 (6): 520–530.

Maguire, G., and M.K. Byrne. 2017. The law is not as blind as it seems: Relative rates of vicarious trauma among lawyers and mental health professionals. *Psychiatry, Psychology and Law* 24 (2): 233–243.

McCaffrey, G. 2019. A humanism for nursing? *Nursing Inquiry* 26 (2): e12281. https://doi.org/10. 1111/nin.12281.

McCann, L., and L.A. Pearlman. 1990. Vicarious traumatisation: a framework for understanding the psychological effects of working with victims. *Journal of Traumatic Stress* 3 (1): 131–149. https://doi.org/10.1111/nin.12281.

Meadors, P., A. Lamson, M. Swanson, M. White, and N. Sira. 2010. Secondary traumatization in pediatric healthcare providers: Compassion fatigue, burnout, and secondary traumatic stress. *OMEGA-Journal of Death and Dying* 60 (2): 103–128. https://doi.org/10.2190/OM.60.2.a.

Mealer, M., J. Jones, J. Newman, K.K. McFann, B. Rothbaum, and M. Moss. 2012. The presence of resilience is associated with a healthier psychological profile in intensive care unit (ICU) nurses: results of a national study. *International Journal of Nursing Studies* 49292–49299. https://doi. org/10.1016/j.ijnurstu.2011.09.015.

Melamed, S., A. Shirom, S. Toker, and I. Shapira. 2006. Burnout and risk of type 2 diabetes: A prospective study of apparently healthy employed persons. *Psychosomatic Medicine* 68 (6): 863–869. https://doi.org/10.1097/01.psy.0000242860.24009.f0.

Munroe, J.F., J. Shay, L. Fisher, C. Makary, K. Rapperport, and R. Zimering. 1995. Preventing compassion fatigue: a team treatment model. In *Compassion fatigue: Coping with secondary traumatic stress disorder in those who treat the traumatized,* ed. C.R. Figley, 209–231. New York, NY: Brunner/Mazel.

Nelson-Gardell, D., and D. Harris, D. 2003. Childhood abuse history, secondary traumatic stress, and child welfare workers. *Child Welfare* 82(1)5–26.

Numminen, O., H. Repo, and H. Leino-Kilpi. 2017. Moral courage in nursing: A concept analysis. *Nursing Ethics* 24 (8): 878–891. https://doi.org/10.1177/0969733016634155.

Oakley, B., A. Knafo, and M. McGrath. 2011. Pathological altruism: An introduction. In *Pathological altruism,* ed. B. Oakley, A. Knafo, G. Madhavan, and D.S. Wilson, 1–9. New York, NY: Oxford University Press.

Oskrochi, Y., M. Maruthappu, M. Henriksson, A.H. Davies, and J. Shalhoub. 2016. Beyond the body: A systematic review of the nonphysical effects of a surgical career. *Surgery* 159 (2): 650–664.

Oxford Dictionary. 2019. *Discernment*. https://www.lexico.com/en/definition/discernment.

Papazoglou, K., M. Koskelainen, and N. Stuewe. 2019. Examining the relationship between personality traits, compassion satisfaction, and compassion fatigue among police officers. *Sage Open* 9 (1): 2158244018825190.

Pike, A.W. 1991. Moral outrage and moral discourse in nurse physician collaboration. *Journal of Professional Nursing* 7 (6): 351–362.

Pines, A.M. 1993. Burnout. In *Handbook of stress—Theoretical and clinical aspects*, 2nd ed, ed. L. Goldberger and S. Breznitz, 386–402. New York, NY: Free Press.

Pines, A.M., and E. Aronson. 1988. *Career burnout: Causes and cures*. New York, NY: Free Press.

Pines, A., and C. Maslach. 1978. Characteristics of staff burnout in mental health settings. *Hospital Community Psychiatry* 29: 233–237. https://doi.org/10.1176/ps.29.4.233.

Rees, C.S., L.J. Breen, L. Cusack, and D. Hegney. 2015. Understanding individual resilience in the workplace: the international collaboration of workforce resilience model. *Frontiers in Psychology* 6: 73. https://doi.org/10.3389/fpsyg.2015.00073.

Reid, L. 2014. Moral distress and moral disorientation in the context of social accountability. *Journal of Graduate Medical Education* 6 (3): 583–584. https://doi.org/10.4300/JGME-D-14-00345.1.

Ross, C.A., N.S. Berry, V. Smye, and E.M. Goldner. 2018. A critical review of knowledge on nurses with problematic substance use: The need to move from individual blame to awareness of structural factors. *Nursing Inquiry* 25 (2): e12215. https://doi.org/10.1111/nin.12215.

Rushton, C.H. 2013. Principled moral outrage: an antidote to moral distress. *AACN Advanced Critical Care* 24 (1): 82–89. https://doi.org/10.1097/NCI.0b013e31827b7746.

Rushton, C.H., A.W. Kaszniak, and J.S. Halifax. 2013. A framework for understanding moral distress among palliative care clinicians. *Journal of Palliative Medicine* 16 (9): 1074–1079. https://doi.org/10.1089/jpm.2012.0490.

Salloum, A., D.C. Kondrat, C. Johnco, and K.R. Olson. 2015. The role of self-care on compassion satisfaction, burnout and secondary trauma among child welfare workers. *Children and Youth Services Review* 49: 54–61. https://doi.org/10.1016/j.childyouth.2014.12.023.

Salston, M.D., and C.R. Figley. 2003. Secondary traumatic stress effects of working with survivors of criminal victimizations. *Journal of Traumatic Stress* 16 (2): 167–174.

Santos, M.C., L. Barros, and E. Carolino. 2010. Occupational stress and coping resources in physiotherapists: a survey of physiotherapists in three general hospitals. *Physiotherapy* 96: 303–310. https://doi.org/10.1016/j.physio.2010.03.001.

Seligman, M.E.P. 2011. *Flourish: A visionary new understanding of happiness and well-being*. New York, NY: Free Press.

Severinsson, E. 2003. Moral stress and burnout: Qualitative content analysis. *Nursing & Health Sciences* 5 (1): 59–66. https://doi.org/10.1046/j.1442-2018.2003.00135.x.

Sexton, L. 1999. Vicarious traumatisation of counsellors and effects on their workplace. *British Journal of Guidance and Counselling* 27 (3): 393–403. https://doi.org/10.1080/030698899082 56279.

Sinclair, S., S. Raffin-Bouchal, L. Venturato, J. Mijovic-Kondejewski, and L. Smith-MacDonald. 2017. Compassion fatigue: A meta-narrative review of the healthcare literature. *International Journal of Nursing Studies* 69: 9–24. https://doi.org/10.1016/j.ijnurstu.2017.01.003.

Storch, J.L., and N. Kenny. 2007. Shared moral work of nurses and physicians. *Nursing Ethics* 14 (4): 478–491. https://doi.org/10.1177/0969733007077882.

Talbot S.G., and W. Dean. 2018. Physicians aren't 'burning out.' They are suffering from moral injury. *STAT*. Retrieved from https://www.statnews.com/2018/07/26/physicians-not-burning-out-they-are-suffering-moral-injury/.

Thompson, F. 2002. Moving from codes of ethics to ethical relationships for midwifery practice. *Nursing Ethics* 9 (5): 522–536.

Vandevala, T., L. Pavey, O. Chelidoni, N.F. Chang, B. Creagh-Brown, and A. Cox. 2017. Psychological rumination and recovery from work in intensive care professionals: Associations with stress, burnout, depression and health. *Journal of Intensive Care* 5 (1): 16. https://doi.org/10.1186/s40 560-017-0209-0.

Viladarga, R., J.B. Luoma, S.C. Hayes, J. Pistorello, M.E. Levin, M.J. Hildebrandt, B.Kohlenberg, N.A. roget, and F. Bond. 2011. Burnout among the addiction counseling workforce: the differential roles of mindfulness and values-based processes and work-site factors. *Journal of Substance Abuse Treatment* 40: 323–335. https://doi.org/10.1016/j.jsat.2010.11.015.

Wachholz, A., G.D.L. Dalmolin, A.M.D. Silva, R. Andolhe, E.L.D. Barlem, and S.B. Cogo. 2019. Moral distress and work satisfaction: what is their relation in nursing work? *Revista da Escola de Enfermagem da USP* 53: e03510.

Wacker, R., and I. Dziobek. 2018. Preventing empathic distress and social stressors at work through nonviolent communication training: A field study with health professionals. *Journal of Occupational Health Psychology* 23 (1): 141–150. https://doi.org/10.1037/ocp0000058.

Zhang, Y.Y., C. Zhang, X.R. Han, W. Li, and Y.L. Wang. 2018. Determinants of compassion satisfaction, compassion fatigue and burn out in nursing: a correlative meta-analysis. *Medicine* 97 (26): e11086. https://doi.org/10.1097/MD.0000000000011086.

Zwisohn, M., W. Handley, D. Winters, and A. Reiter. 2018. Vicarious trauma in public service lawyering: How chronic exposure to trauma affects the brain and body. *Richmond Public Interest Law Review* 22: 101.

Chapter 3
A Socio-Ecological Perspective

Abstract One way to explore a complex concept is to consider it as layered and interactional and seek to understand it from a variety of perspectives. Here, I employ a socio-ecological model to consider the nested nature of various aspects of this concept. Moral distress is a phenomenon with potentially dynamic effects from micro-level concerns for health care providers as individuals, both personally and professionally, through to macro-level concerns for political and health care systems. Further, moral distress experiences occur over time and may also occur in virtual realities. These multiple and sometimes co-occurring influences are difficult to account for, may not always be recognized as factors in ethical decision making, and are even more difficult to address without an understanding of the interconnections that may be involved.

Keywords Moral stress · Socio-ecological model · Costs · Consequences · Benefits

3.1 Introduction

It could be argued that to focus this book on nursing, with support from multidisciplinary sources of research, is too restrictive. However, according to the World Health Organization (2006) nurses represent the largest group of professional providers in health care. Thus exploring moral distress through a nursing first perspective and then beyond is a good place to start. The problem is that nurses and other health care providers often suffer with experiences of moral distress related to ethical issues in their practice. The problem is not that moral distress experiences happen, rather it is in the response of suffering because suffering eventually needs an outlet. Further, prolonged suffering has costs and consequences. This matters because of alarmingly high rates in health care of compassion fatigue, burnout, job turnover, and sometimes attrition from one's profession entirely. Any and all of these circumstances ultimately impact quality patient care and positive health outcomes.

As a concept that has been studied for almost four decades, moral distress is, surprisingly, still up for discussion and debate. The argument is largely about where the borders should be that distinguish moral distress from general work-related stress on one end and post traumatic stress disorder (PTSD) on the other. These borders are important, beyond just an exercise in academic navel gazing on a theoretical concept, because an accurate definition and valid tools for measurement would help to identify the actual scale of moral distress, as well as allow for formal evaluation of a variety of interventions. However, there will not be a one-size-fits-all, single solution.

What we know is that moral distress exists as real, complex, and difficult personal experiences among many types of health care providers and in a variety of health care settings. What we don't know much about are the factors that would most effectively prevent some experiences of moral distress and promote competence and confidence to navigate ethics in practice. We do have some ideas about what might work, but the follow through is not always there to sustain these initiatives long enough to adequately evaluate them.

To begin, the work of nurses (and many caregiving professionals) should be expected to have a complex matrix of anticipated stressors. Caring work is largely around the clock (24-hours-a-day/7 days-a-week) and year round. Shift work alone is known to be a significant stressor on both the body and the mind. This work often situates professionals to be in close proximity to the pain, suffering, and vulnerability of others with unknown outcomes and diverse variables (Peter and Liaschenko 2004). This in turn may uncover one's own pain, suffering, and innate vulnerabilities making health care practice a unique moral arena, at times rife with ethical land mines.

Working with individuals, families and/or communities, in crisis or in difficult situations, is a naturally stressful experience. In the human body, stress is linked to the release of adrenaline, norephinephrine, and cortisol. This is hugely advantageous in an emergency situation, but an ongoing stress response can be detrimental to our physical and emotional health and well-being. So why would anyone choose to do work that is likely to bring on a sustained stress response? Well, this is the brave work of carers and there are varied reasons why people come into caring work, stay in it, and also leave. It is the privileged work of those who have the potential to make a difference; the honour to be part of something larger than themselves.

3.2   Moral Stress

Stress can be defined as an event that strains or exceeds the resources of an individual or a system (Lützén and Ewalds-Kvist 2013). Stress in life cannot be completely avoided and in fact it should not be. From a strengths-based approach, stress can be considered as an important catalyst for change and growth. However, even eustress, stress that we judge to be positive (e.g., getting married, having a baby) impacts our physiology and induces a stress response with hormone release, immune system responses, and inflammation. A chronic state of responding to stress is detrimental to health and well-being and can damage the body and the mind.

Swedish nursing professor, Kim Lützén and her colleagues prefer to use the term moral stress to describe moral distress-like responses with more physiological components. In their study, moral stress is explored as a occupational health risk with potential long-term consequences for caring professionals that stems from situations of low personal control (perceived constraints) and high moral demands that occur in or near a client's private sphere (Lützén et al. 2003). Moral stress is defined as a phenomenon that results from health care providers possessing a sensitivity for moral issues that manifests as intuitional knowing that something should occur (Lützén et al. 2000, 2006).

Research on moral stress and moral distress has explored routine moral tensions and burdens in nursing practice and considers these as moral demands that are normative for the occupation (Lützén and Kvist 2012; Cribb 2011; Lützén et al. 2003). Moral stress occurs on a spectrum that may have both positive and negative outcomes as nurses negotiate ethical conflicts in morally demanding situations.

In a Dutch multi-sector study with over 350 nurses, morally distressing caregiving experiences were linked to factors such as complying with family's wishes (versus patient's wishes), physician/nurse disagreements, and unsafe staffing levels (de Veer et al. 2013). (See Table 3.1 for common factors of moral distress experiences across many studies.) Further, a lack of time for patient care and a lack of time to share concerns with colleagues were highlighted as key sources of moral distress. Protective

Table 3.1 Common factors in moral distress experiences

Shared factors	Shared goals	Shared barriers
Time	Quality patient care	Unable to support colleagues (e.g., listening, validating, advising) Poor information gathering/sharing (e.g., documentation)
Power (powerlessness; real or perceived)	Quality patient care	Hierarchies Lack of leadership support or unsupportive style of leadership Policies that conflict with care Part-time or casual employment status
Relationship and communication	Quality patient care	Physician and nurse Nurse and other members of health care team Nurse and organizational leadership Health care team and patient Health care team and family Health care team and organization
Staffing and workload	Quality patient care	Legal and/or professional risks Working outside of one's scope of practice Other provider's incompetency Work overload Absenteeism and presenteeism

factors include relational leadership styles and full-time status. This study reiterates the consequences of unresolved moral distress experiences, as found in many studies, to be keenly related to job dissatisfaction, early retirement, and/or job turnover.

Therefore, moral distress begins with an individual's moment of awareness that something should be happening that isn't or something is happening that should not be. It is potentiated by situations of perceived powerlessness and past/present/future patient risk for harm. It involves our work with others, as both contributing to shared goals and to encountering barriers to achieving those goals. It often involves a time crunch. It may negatively impact an individual's sense of self-efficacy, confidence, work satisfaction, and decision to stay in a workplace or a profession.

In an attempt to measure the negative stress of moral issues in health care practice, Swedish scholar Ann-Louise Glasberg et al. (2006) constructed and validated a tool known as the Stress of Conscience Questionnaire (SCQ) with over four hundred health care providers. Two factors associated with 'troubled conscience' were identified with this scale: (1) internal demands; and, (2) external demands/restrictions. And so we begin with at least two layers to consider.

3.3   Conceptual Framework

The original concept of a socio-ecological model was developed in the 1970s by Urie Bronfenbrenner, a Russian-born developmental psychologist with military experience. In the context of a socio-ecological theory, no single system works independently of the other systems; much like our body. The parts must be considered in relation to the whole. Therefore, moral distress understandings and interventions that soley focus on the individual nurse and thoughts/behaviours/feelings cannot be a sufficient solution to the problem. However, individually focused strategies are not completely irrelevant either. Every change has the potential for a ripple effect in the overall system; an interdependent benefit or burden. A number of authors support the notion that moral distress must be understood from an organizational, structural, and systemic perspective and that an individual perspective deflects important questions and answers (Austin 2012; Varcoe et al. 2012). Therefore various contexts must be considered in order to gain perspective of the bigger picture. Further discussed will be an explanation of the concept of moral distress using a systems approach from time and space through to the core of the individual experience (see Fig. 3.1).

3.4   In Consideration of Time

If there is one barrier, one challenge, one resource in dire need in contemporary health care environments and among health care providers it is time. The consensus is, there is 'not enough time'. Providers feel that they 'cannot take the time'. Some try to 'make time' for empathetic gestures and a few kind words. Again and again we

Fig. 3.1 A socio-ecological
model of moral distress

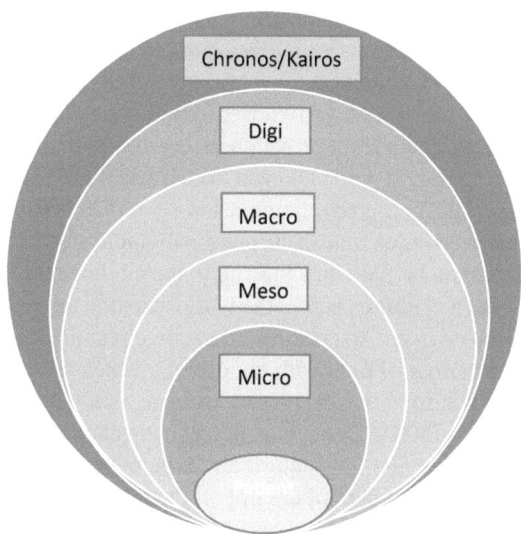

encounter 'ideologies of scarcity' in health care and time, as a key factor, is maybe the foremost example. There has always been the same amount of time in a day; 24-hours to be exact. How have we come to believe that, somehow, now there is less of it?

Chronos is the Greek personification of time as it is considered by the modern western world. Time is in the past, present, and the future. It is linear and can be quantified and calculated. It is chronological and thus events occur in sequence to each other. In contrast, kairos, is about a 'right moment' and has an intangible, qualitative nature to it that can be grasped but not fully captured. Granted, not every culture on Earth views time in such ways, but this is a jumping off point to get started.

There are many metaphors for time and also ways of referring to it that may influence how health care providers think about the time they have and how to use it, such as:

- *Time heals all wounds.*
- *Chasing time.*
- *Time flies.*
- *Time is money.*
- *Time's up.*
- *All the time.*

As a pedagogical approach to nursing education, I have recently adopted the practice of paying more attention to time in the context of sharing historical perspectives with students to ground them in how we have arrived at the present day/moment and to help them to explore where we may be headed in the future. This approach provides valuable context. There are important lessons that have been learned 'over time' that should be passed on to future generations. Unfortunately, humans as a species tend to make many of the same mistakes again and again across each generation, particularly in our interactions with other humans.

Individual perspectives of time and time passing can also be context dependent. Try to recall how you viewed time as a child. Did many activities require endless waiting and a level of patience that you perhaps did not possess? Consider now an average day at work. Did you 'lose track of time' and miss your lunch or coffee break? Do you often arrive home asking yourself what happened to the day? Therefore, perspectives on time in the present moment are not fixed and equal for each individual. Time can be fluid and specific to the circumstances.

Many cultures across the world see time quite differently. In teachings that I have received from First Nation elders, time is considered in a very broad way. One cannot trap or control time with the use of a watch or an alarm clock. Being 'on time' may be when someone decides to arrive at a particular location versus a set time such as 1 pm or 1300 h. Being late, not arriving at all, or arriving uninvited may be culturally acceptable and not require any type of apology, acknowledgment, or explanation. Time is not of paramount importance. Time is eternal. It is iterative, continuous, and even with death, time does not stop. It continues on through many generations.

Military cultures traditionally view time as a priority and with precision. Health care, with its roots in military service and religious orders, often occurs in environments that have rigid and established routines firmly embedded in timely responses and tasks. In the context of a hospital, as an example, the set times for vital signs, medication administration, meals, and visiting may take precedence over the needs of individual patients and families, even in the modern era of patient-family centred care. Some structure is obviously required for an institution to function, however I argue that unquestioned adherence to long-held, Eurocentric views about 'time' can limit health care providers' perspectives on ethical situations and their ability to navigate ethical decision making in their practice. Further to this, there is usually little time allocated to building capacity for ethics in practice. If only there was time for that!

With time viewed as a valuable commodity, health care providers make a myriad of decisions every day about what they have time for, what is worth their time, and how their time will best be spent. For some patients, 'time is all they have left' with many of their past roles and responsibilities having been stripped away by illness and disease. For other patients, 'time runs short' and decisions are required quickly before death arrives. In these cases, hesitation, non-action, or the wrong choice cannot be undone. Time as a patient stops at death and then there is only the shadow of past moments to reflect on for those who remain. 'Lost time' may be responded to with grief, shame, frustration, and/or anger.

Time is also in the future. Technology is supposed to 'save time' for health care providers with the use of computers, algorithms, artificial intelligence, and robots. What will health care providers do with all that new found time? Will this bring nurses the opportunity to be present with their patients in ways that they are currently, rarely able to do? Or will the future be filled with more missed opportunities for human connection and result in further distancing in high tech(nology), low touch health care environments? These are some of the moral matters of chronos and kairos.

3.5 Space and the Digital Worlds

Health care places and spaces are moving further and further into digital realms. Electronic medical records, virtual provider-patient visits via distance technology, text messages for appointment reminders, and avatar therapy counselling (Craig et al. 2018), are a few examples of this. Health care organizations connect with patients and families through websites and commercials on television, asking for financial gifts through donations and legacy planning. Smiling health care providers working one-on-one with a content patient, who is perhaps expressing their gratitude, is the common media image of our contemporary service delivery of health care. Health care professions students, patients and family members seek further health information and second opinions from YouTube videos, downloadable apps, and Google searches. Social media helps to link brave patients' stories to go-fund-me opportunities and to find support from compassionate others from around the world. Even health care occurs in the context of a global village. Health and health care knowledge is shared, for the most part, globally in our contemporary world, as well as in phenomena like health care tourism. Health care tourism (Karuppan and Karuppan 2010) takes place when and where people can afford it, in countries that offer better, quicker, easier, or otherwise unavailable access to medicines, treatments, or diagnostic tests for the privileged few.

Patients' health care experiences, good, bad, and everything in between, are often shared on social media with tweets on Twitter, pictures on SnapChat, or a rant on Facebook. However, health care providers must beware. Even as a patient or family member, nurses have been publically disciplined for speaking out and speaking up in digital domains. A current case is under the highest level of appeal in Saskatchewan, Canada. This is the case of registered nurse Carolyn Strom and a charge of professional misconduct (Taylor 2019). Found guilty of the charge, she was fined $26,000 for social media comments she posted in 2015. What did she do? She criticized the care (generally) that her grandfather had received at end-of-life in an institution (not her workplace) and identified herself as a nurse in the posting. The debate of her wrongdoing is ongoing and the divergent opinions are that her actions were unprofessional or that her actions should be upheld as freedom of speech. The final decision still has not been made or at least has not been made publically available.

This brings to light the need for 'safe' spaces and places for health care providers to engage in conversations, discussions, or debates about ethics in practice. Clearly for nurse Carolyn, social media was not a safe space for her to be open and honest about her experiences of being a granddaughter with a beloved family member receiving palliative care. Her expression of the message '*I'm a nurse and I know what I'm talking about when I see poor care practices*' was deemed by her regulatory college to be a serious misjudgment and a mistake that she would pay for both personally and professionally.

Many nurses are afraid of the potential repercussions of being accused of breaching standards of practice and/or codes of ethics. There are serious and real risks. The costs are high and more than just financial. Disciplinary processes can take years to unwind in both private and public spaces. In Canada, the disciplinary process for nurses allows documents and content to be made publically available online. There is a social factor here, public shaming if you will. This particular case has been all over the Canadian media for many years. At times, guilt can appear to be assumed with the need to prove innocence squarely on the shoulders of the nurse. All in the name of quality care and protection of the public's safety. Therefore, health care providers are often wary to enter into discussions about ethics in their practice at all. There are many risks, both personal and professional that may need to be weighed and navigated. Even bringing an everyday ethical issue forward, such as to an ethics committee or bioethicist, can feel like an act of whistleblowing (Blenkinsopp et al. 2019; Pohjanoksa et al. 2019). And where would one even begin to address a bigger ethical issue?

From an Aboriginal perspective and my (humbly limited) understanding of traditional teachings, there is also a need to be inclusive of nature and spiritual connections for the ethics of our everyday activities in places and spaces. This may include considering elders (past, present and future), animals, and non-living beings (such as rocks, mountains, etc.) in ethical decision making processes. There is beautiful work by Indigenous (Mohawk) scholar Marlene Brant Castellano (2004) on contextualizing Aboriginal ethics. She uses the metaphor of a deciduous tree that I always turn to. In this writing, she describes our individual behaviours as the leaves on the tree; protocols, customs, ethics, and rules as the branches; our values and deepest beliefs as the trunk; the roots just beneath the earth as our world view; and, the deepest roots as our connection to the world of spirit. This is a simple metaphor with room for complexity, spirituality, and rich discussions that can help us to re-consider contexts of space and place, in digital worlds and beyond.

3.6   Macro Systems

Macro system considerations occur at the level of health care systems and political agendas in this particular model. Of primary importance here is the worldwide nursing shortage. According to the American Association of Colleges of Nursing (2019), the current nursing shortage is estimated to continue until at least 2030 due to high retirement and turnover rates, a shortage of doctoral prepared nursing faculty, decreased job satisfaction, and increased stress levels. The loss of seasoned nurses from the global health care system is not just about numbers; it is also the loss of wisdom that is likely the glue holding together broken systems.

The Canadian health care system is still mostly physician directed with largely publicly funded health services with some privatization. As with education, health care decisions are largely made at the provincial level and therefore there are differences in each province or territory across the country. Changes to actually address factors that contribute to moral distress experiences for nurses and other health care providers would require a major systems overhaul. Nurses may be in a valuable position to recommend where and when those changes should occur.

Unfortunately, and I will give the example of the situation in my home province of Ontario, registered nurses' positions are changing. Due to efforts to minimize costs and maximize 'efficiencies' nurses are often not replaced upon retirement or are replaced with differently qualified care providers, some who may be professionally unregulated. I am mindful to acknowledge here that there is absolute value to having a variety of qualifications, expertise, and knowledge on a team. However, research shows that the presence of registered nurses (RNs) makes a difference that should be part of decision making equations. In a study by American nurse researcher Linda Aiken et al. (2017), the staff mix of 243 European hospitals across six countries was explored through surveys of patients and nurses, and hospital data. Skilled professional nurses were associated with lower patient mortality, higher ratings on care from patients, and better patient outcomes. Statistics from studies like this should be used to set provincial government and ministry of health standards for optimized staffing levels that reflect evidence-informed facts on best practices for quality patient care.

3.7 Meso Systems

Meso systems occur at organizational and team levels in this particular model. Research conducted almost 25 years ago identified organizational issues that lead to considerable moral tensions for health care providers, including lack of supervisory support, limited professional development opportunities, low influence, and high workloads (Severinsson and Kamaker 1997). This information remains relevant today and impacts from these tensions continue to influence ethical climate within organizations.

Ethical climate can be defined as a shared understanding of individual and collective perspectives about ethical behaviour (e.g., practices, procedures, policies) and has been empirically linked with experiences of moral distress, job turnover, work satisfaction, well-being and quality patient care (Koskenvuori et al. 2019). Ethical climate sends a message to health care providers about whether or not an organization's mission, vision, and values are being lived or are just words on a page or website. It shows employees whether or not they need to follow policy and procedure or need to learn the unwritten rules and hidden agendas. Ethical climate is expressed through the people and behaviours that are rewarded and upheld in an organization.

Ethical climate speaks to the things that are prioritized in an organization. Many health care organizations feel pressure to comply to set standards to maintain their funding, avoid fines, and/or obtain accreditation. Measures such as emergency room wait times, surgical waiting lists, handwashing compliance, infection rates, and readmission rates, as a few examples, can become priorities. These should appear as odd measures of caring for and about people, but are often accepted as stand alone measures of quality care.

The prioritizing of numbers over people can leave ethics in practice largely unsupported and can contribute to an environment with high rates of absenteeism due to sickness or injuries and difficulty with recruitment and retention (e.g., unfilled positions, job turnover). This is the making of an unstable team environment.

The most vital meso-level effects are those involving negative outcomes for the quality of relationships amongst health care professionals, health care provider-patient-and-family relationships, as well as on the level of engagement in meeting professional obligations, ethically and competently. Moral distress experiences can impact relationships between nurses and their colleagues and have an effect on effective multidisciplinary team functioning. This can manifest in untoward patient and family care outcomes and experiences. These are costs and consequences that cannot be readily measured.

In terms of a strong ethical climate in an organization and good support among colleagues and from supervisors, it can be equally difficult to measure the savings and benefits that occur and to fully articulate the value of a well functioning organization and team. Why would you need to? Sometimes there is a need to advocate for not disrupting what is actually working well.

There are excellent models of high functioning organizational cultures with research to back their tried and true strategies. The American Nurses Association offers magnet recognition status for organizations worldwide who wish to certify their commitment to the highest quality of patient care through attention to nurses and their role in the organization (American Nurses Credentialing Centre n.d.). For example, an integrative review of hospitals with magnet hospital designation revealed positive impacts on nursing practice environments such as increased autonomy and sense of control, better relationships, and supportive leadership (Anderson et al. 2018). Further, adopting a shared governance model led to further empowerment of nurses to actively participate and engage with the organization. The trending data shows that what is good for nurses is also good for patients.

For many years, individual sectors of health care (e.g., acute care, community care, public health) were referred to in terms of 'silos'. Health care providers worked in their 'silo' without much interaction with other siloed sectors. Now when I use that term in class, students give me an odd look. Perhaps it is because a silo is not a common term to bring to mind or maybe it is because today's health care environments need to be far more integrated than in the past. Nurses and other health care providers routinely work with other (outside) organizations. Working with other health care organizations is no surprise, but the scope of health care has expanded to often include schools and daycares, community policing, occupational health, social services, financial services, counselling and addictions services, multicultural associations, and non-profit organizations, as just a few examples. Opportunities for formal partnerships and informal collaborations can be vital, especially for organizations that have limited funding and/or resources on their own. While there can be many benefits from finding synergies between organizations and teams, this kind of work also takes time and energy dedicated to establishing and maintaining positive relationships.

However, toxic workplaces do exist and have been identified in health care environments. A study of hundreds of workplaces reveals three common key aspects of 'sick organizations' and these include: (1) sick systems; (2) sick leaders; and,

(3) dysfunctional colleagues (White and Schoonover-Shoffner 2016). Contributing factors include poor communication, non-adherence to policies, lack of defined responsibilities, leaders who are narcissistic, and colleagues who do not accept their own responsibilities.

In a study of 120 nurses in Iran, moral distress and organizational support were surveyed using the MDS and the Survey of Perceived Organizational Support (SPOS; evaluates an employee's beliefs about: their organization's commitment to them personally; if the organization values their continued membership; and, if the organization values their well-being) (Robaee et al. 2018). In this study, moral distress was found to be high and organizational support was low. There was no statistically significant correlation between the two. This speaks to the need for better measurement tools or the problematic issue of self-reporting where nurses may not feel safe to honestly report dissatisfaction with their employer.

Finally, nursing, itself, has a bad reputation; the metaphor of nurses 'eating their young' referring to how nursing students often experience horizontal violence (e.g., emotional or verbal abuse) during their clinical placements or as a new graduate at the hands of other nurses (Longo 2007). A study of workplace bullying among over 100 nurse leaders found that 60% reported experiencing intense and persistent exposure to bullying (Hampton e al. 2019). It is difficult for an individual to rise above moral distress in a culture of incivility and insubordination.

3.8 Micro Systems

Micro systems occur at the individual level (intrapersonal) and between individuals (interpersonal) in this particular model. While moral distress can sensitize individuals to important ethical issues in practice (Austin et al. 2005), eventually, a nurse's sense of accountability to the patient, responsibility for the outcome, and perceived obligations will require a response. Responses may include avoidance behaviours (e.g., depersonalization), denial, or surrender (Epstein and Hamric 2009). Studies have correlated nurses' experiences of moral distress with leaving a job (such as Corley 1995; Corley et al. 2001; Hamric and Blackhall 2007) as well as with the phenomenon of burnout (such as Gustafsson et al. 2010; Meltzer and Huckabay 2004; Sundin-Huard and Fahy 1999). Decreased job satisfaction may also become an issue for successful nurse retention (Severinsson and Hummelvoll 2001) within a particular job or within the profession itself, in the context of moral distress experiences and poor team cohesion.

Broken systems, unhealthy organizations, and uncooperative teams can lead to unhealthy and unhappy employees. One study that explored sickness presenteeism and sickness absenteeism found that exhaustion and excessive job demands were associated with both (Brborović et al. 2017). Further, burnout and depersonalization were outcomes of sickness presenteeism over time. There are both measured and unmeasured costs to individuals, their colleagues, their workplaces, and the patients and families in their care. The outcome is a nurse who may feel haunted, tormented, shattered, and/or crushed by their morally distressing experiences of caring. This is a terrible reality. However, American nursing scholar Betty Ramber et al. (2010) write about the possibility of moral eustress (eu-meaning good) and its potential contribution to moral growth and development; there is room for hope.

3.9 Conclusion

Bringing this socio-ecological model back to centre, among health care professionals the shared goal should always be for the good of the patient. Good care has three core components: (1) caring relationships; (2) caring behaviours; and, (3) holistic approaches that consider all domains (e.g., physical, emotional, social, spiritual, etc.) (Gastmans et al. 1998). Care and care-full approaches can and should happen at all the levels of all the systems.

As the effects of moral distress experiences ripple out from an individual health care provider, to other members of the health care team, and toward patients and families, the overall health care system itself is also affected (Clarke et al. 2001; Kälvemark et al. 2004). Given the potential for adverse, multi-level impacts, as discussed here, moral distress must continue to be considered a critical issue in contemporary health care. However, the cumulative influences of macro, meso, micro, time, and space effects of moral distress are difficult to objectively and accurately understand, measure, and evaluate. Therefore, it can be tricky to fully articulate the problems and this makes advocating for the value of addressing moral distress in practice quite challenging. And furthermore, this measurability issue affects the potential for accurately evaluating interventions designed to address, resolve, or prevent moral distress. Thus it is necessary to continue to explore and deepen our understanding of moral distress in all contexts of health care delivery. With that in mind, it might be helpful to take a closer look at what's happening in the silos.

References

Aiken, L.H., D. Sloane, P. Griffiths, A.M. Rafferty, L. Bruyneel, M. McHugh, C.B. Maier, T. Moreno-Casbas, J.E. Ball, D. Ausserhofer, and W. Sermeus. 2017. Nursing skill mix in European hospitals: Cross-sectional study of the association with mortality, patient ratings, and quality of care. *BMJ Quality & Safety* 26 (7): 559–568. https://doi.org/10.1136/bmjqs-2016-005567.

American Association of Colleges of Nursing. 2019. *Facts sheets: Nursing shortage*. Retrieved from https://www.aacnnursing.org/News-Information/Fact-Sheets/Nursing-Shortage

American Nurses Credentialing Centre. n.d. *Magnet recognition program*. Retrieved from https://www.nursingworld.org/organizational-programs/magnet/

Anderson, V.L., A.N. Johnston, D. Massey, and A. Bamford-Wade. 2018. Impact of MAGNET hospital designation on nursing culture: An integrative review. *Contemporary Nurse* 54 (4–5): 483–510. https://doi.org/10.1080/10376178.2018.1507677.

Austin, W., G. Lemermeyer, L. Goldberg, V. Bergum, and M. Johnson. 2005. Moral distress in healthcare practice: The situation of nurses. *HEC Forum* 17 (1): 33–48.

Austin, W. 2012. Moral distress and the contemporary plight of health professionals. *HEC Forum* 24 (1): 27–38. https://doi.org/10.1007/s10730-012-9179-8.

Blenkinsopp, J., N. Snowden, R. Mannion, M. Powell, H. Davies, R. Millar, and J. McHale. 2019. Whistleblowing over patient safety and care quality: a review of the literature. *Journal of Health Organization and Management* 33 (6): 737–756. https://doi.org/10.1108/JHOM-12-2018-0363.

Brborović, H., Q. Daka, K. Dakaj, and O. Brborović. 2017. Antecedents and associations of sickness presenteeism and sickness absenteeism in nurses: A systematic review. *International Journal of Nursing Practice* 23 (6): e12598. https://doi.org/10.1111/ijn.12598.

Castellano, M.B. 2004. Ethics of Aboriginal research. *Journal of Aboriginal Health* 1: 98–114.

Clarke, H., H. Spence Laschinger, P. Giovannetti, J. Shamian, D. Thomson, and A. Tourangeau. 2001. Nursing shortages: Workplace environments are essential to the solution. *Hospital Quarterly* 4 (4): 50–58. https://doi.org/10.12927/hcq..17434

Corley, M.C. 1995. Moral distress of critical care nurses. *American Journal of Critical Care* 4 (4): 280–285. https://doi.org/10.4037/ajcc1995.4.4280.

Corley, M.C., R.K. Elswick, M. Gorman, and T. Clor. 2001. Development and evaluation of a moral distress scale. *Journal of Advanced Nursing* 33 (2): 250–256. https://doi.org/10.1111/j.1365-2648.2001.01658.x.

Craig, T.K., M. Rus-Calafell, T. Ward, J.P. Leff, M. Huckvale, E. Howarth, R. Emsley, and P.A. Garety. 2018. Avatar therapy for auditory verbal hallucinations in people with psychosis: a single-blind, randomised controlled trial. *The Lancet Psychiatry* 5 (1): 31–40. https://doi.org/10.1016/S2215-0366(17)30427-3.

Cribb, A. 2011. Integrity at work: Managing routine moral stress in professional roles. *Nursing Philosophy* 12: 119–127. https://doi.org/10.1111/j.1466-769X.2011.00484.x.

de Veer, A.J., A.L. Francke, A. Struijs, and D.L. Willems. 2013. Determinants of moral distress in daily nursing practice: A cross sectional correlational questionnaire survey. *International Journal of Nursing Studies* 50 (1): 100–108. https://doi.org/10.1016/j.ijnurstu.2012.08.017.

Epstein, E.G., and A.B. Hamric. 2009. Moral distress, moral residue, and the crescendo effect. *The Journal of Clinical Ethics* 20 (4): 330–342.

Gastmans, C., P. Schotsmans, and B. Dierckx de Casterle. 1998. Nursing considered a moral practice: A philosophical-ethical interpretation of nursing. *Kennedy Institute of Ethics Journal* 8 (1): 43–69. https://doi.org/10.1353/ken.1998.0002.

Glasberg, A.L., S. Eriksson, V. Dahlqvist, E. Lindahl, G. Strandberg, A. Söderberg, V. Sørlie, and A. Norberg. 2006. Development and initial validation of the stress of conscience questionnaire. *Nursing Ethics* 13 (6): 633–648. https://doi.org/10.1177/0969733006069698.

Gustafsson, G., S. Eriksson, G. Strandberg, and A. Norberg. 2010. Burnout and perceptions of conscience among health care personel: A pilot study. *Nursing Ethics* 17 (1): 23–38. https://doi.org/10.1177/0969733009351950.

Hampton, D., K. Tharp-Barrie, and M. Kay Rayens. 2019. Experience of nursing leaders with workplace bullying and how to best cope. *Journal of Nursing Management* 27 (3): 517–526. https://doi.org/10.1111/jonm.12706.

Hamric, A.B., and L.J. Blackhall. 2007. Nurse-physician perspectives on the care of dying patients in intensive care units: Collaboration, moral distress, and ethical climate. *Critical Care Medicine* 35 (2): 422–429. https://doi.org/10.1097/01.CCM.0000254722.50608.2D.

Kälvemark, S., A.T. Höglund, M.G. Hansson, P. Westerholm, and B. Arnetz. 2004. Living with conflicts-ethical dilemmas and moral distress in the health care system. *Social Science & Medicine* 58 (6): 1075–1084. https://doi.org/10.1016/S0277-9536(03)00279-X.

Karuppan, C.M., and M. Karuppan. 2010. Changing trends in health care tourism. *The Health Care Manager* 29 (4): 349–358. https://doi.org/10.1097/HCM.0b013e3181fa05f9.

Koskenvuori, J., O. Numminen, and R. Suhonen. 2019. Ethical climate in nursing environment: a scoping review. *Nursing Ethics* 26 (2): 327–345. https://doi.org/10.1177/0969733017712081.

Longo, J. 2007. Horizontal violence among nursing students. *Archives of Psychiatric Nursing* 21 (3): 177–178. https://doi.org/10.1016/j.apnu.2007.02.005.

Lützén, K., A. Cronqvist, A. Magnusson, and L. Andersson. 2003. Moral stress: Synthesis of a concept. *Nursing Ethics* 10 (3): 312–322. https://doi.org/10.1191/0969733003ne608oa.

Lützén, K., V. Dahlqvist, S. Eriksson, and A. Norberg. 2006. Developing the concept of moral sensitivity in health care practice. *Nursing Ethics* 13 (2): 187–196. https://doi.org/10.1191/096 9733006ne837oa.

Lützén, K., and B. Ewalds-Kvist. 2013. Moral distress and its interconnection with moral sensitivity and moral resilience: Viewed from the philosophy of Viktor E Frankl. *Journal of Bioethical Inquiry* 10 (3): 317–324. https://doi.org/10.1007/s11673-013-9469-0.

Lützén, K., A. Johansson, and G. Nordstrom. 2000. Moral sensitivity: Some differences between nurses and physicians. *Nursing Ethics* 7 (6): 520–530.

Lützén, K., and B.E. Kvist. 2012. Moral distress: A comparative analysis of theoretical understandings and inter-related concepts. *HEC Forum* 24 (1): 13–25.

Maslach, C., and S.E. Jackson. 1981. The measurement of experienced burnout. *Journal of Organizational Behavior* 2 (2): 99–113. https://doi.org/10.1002/job.4030020205.

Meltzer, L., and L. Huckabay. 2004. Critical care nurses' perceptions of futile care and its effect on burnout. *American Journal of Critical Care* 13 (3): 202–208. https://doi.org/10.4037/ajcc2004. 13.3.202.

Peter, E., and J. Liaschenko. 2004. Perils of proximity: a spatiotemporal analysis of moral distress and moral ambiguity. *Nursing Inquiry* 11 (4): 218–225. https://doi.org/10.1111/j.1440-1800. 2004.00236.x.

Pohjanoksa, J., M. Stolt, R. Suhonen, and H. Leino-Kilpi. 2019. Wrongdoing and whistleblowing in health care. *Journal of Advanced Nursing* 75 (7): 1504–1517. https://doi.org/10.1111/jan.13979.

Ramber, B., C. Vallett, J.A. Cohen, and J. Tarule. 2010. The moral cascade: Distress, eustress, and the virtuous organization. *Journal of Organizational Moral Psychology* 1 (1): 41–54.

Robaee, N., F. Atashzadeh-Shoorideh, T. Ashktorab, A. Baghestani, and M. Barkhordari-Sharifabad. 2018. Perceived organizational support and moral distress among nurses. *BMC Nursing* 17 (1): 2. https://doi.org/10.1186/s12912-017-0270-y.

Severinsson, E., and J.K. Hummelvoll. 2001. Factors influencing job satisfaction and ethical dilemmas in acute psychiatric care. *Nursing and Health Sciences* 3 (2): 81–90. https://doi.org/ 10.1046/j.1442-2018.2011.00076.x.

Severinsson, E., and D. Kamaker. 1997. Clinical nursing supervision in the workplace—Effects on moral stress and job satisfaction. *Journal of Nurse Management* 7: 81–90. https://doi.org/10. 1046/j.1365-2834.1999.00106.x.

Sundin-Huard, D., and K. Fahy. 1999. Moral distress, advocacy and burnout: Theorizing the relationships. *International Journal of Nursing Practice* 5: 8–13. https://doi.org/10.1046/j.1440-172x. 1999.00143.x.

Taylor, S. 2019. Saskatchewan nurse appealing $25K fine over Facebook comments. *The Canadian Press*. Retrieved from https://globalnews.ca/news/5912275/saskatchewan-nurse-court-app eal-fine-facebook/

Varcoe, C., B. Pauly, G. Webster, and J. Storch, J. 2012. Moral distress: Tensions as springboards for action. *HEC Forum* 24 (1):51–62. https://doi.org/10.1007/s10730-012-9180-2

White, P.E., and K. Schoonover-Shoffner. 2016. Surviving (even thriving?) in a toxic workplace. *Journal of Christian Nursing* 33 (3): 142–149. https://doi.org/10.1097/CNJ.0000000000000289

World Health Organization (WHO). 2006. *The world health report 2006: Working together for health*. Geneva: WHO.

Chapter 4
Acute Care Contexts

Abstract Since the term moral distress was first noted almost four decades ago, moral distress experiences have been most thoroughly explored in various acute care settings. The concept of futile or non-beneficial treatments that contribute to unnecessary suffering, especially during palliative and end-of-life care have been found to be a key source of moral distress in acute care settings. Critical/intensive care for adults and children, as well as oncology settings appear to be leaders of moral distress research and exploration. Also explored will be moral distress in burn units, emergency rooms, medical/surgical units, mental health units, operating rooms, perinatal intensive care, renal units, and palliative care/hospice settings. This chapter will contribute to our understanding of the concept of moral distress with studies conducted in acute care hospital settings.

Keywords Acute care · Moral distress · Values · Death · Technology

4.1 Introduction

The first attempts to accurately capture the frequency and intensity of moral distress experiences of nurses were carried out in American intensive/critical care settings in the mid to late 90s (Corley 1995). Despite these efforts of scholarship and inquiry, there still remains a dire need to address moral distress in practice in critical care settings (Anderson 2013). As we continue the search for a broader understanding of moral distress, it is important to consider all the pieces of the puzzle, including contextual factors specific to acute care settings that may set the foundation for moral distress experiences. Certainly the high intensity of stress found in the work of caring for patients and families in acute care settings has a role, where patients are likely to be quite ill and family members may be overwhelmed with concerns and questions. A common thread across the ethics literature is the tension often present between nursing and medicine; a historical and hierarchical power dynamic. However, while dialogue can begin here it cannot end here. Moral distress is not strictly a physician versus nurse issue. There are many other factors at play.

© Springer Nature Switzerland AG 2020
K. Jones-Bonofiglio, *Health Care Ethics through the Lens of Moral Distress*,
The International Library of Bioethics 82,
https://doi.org/10.1007/978-3-030-56156-7_4

4.2   Moral Distress Experiences in Acute Care

Reviewing the recent literature on moral distress in acute care settings (e.g., specific units in hospitals) is foundational two reasons: (1) to identify the spectrum of specialty areas in acute care in which researchers have identified moral distress experiences; and, (2) to highlight potential contemporary variables and common themes that appear to correlate with nurses' and other health care providers' moral distress experiences. Many specialty areas of practice, within acute care settings, are linked to studies on moral distress (see Table 4.1). Clearly, the academic literature supports the ongoing presence of the phenomenon of moral distress among nurses and other health care providers in acute care settings.

4.3   Common Denominators

Studies have used the Moral Distress Scale (MDS; Corley 1995) or a revised version of it with the Psychological Empowerment Instrument (PEI) (such as Browning 2013), the Ethical Work Environment Survey (such as Corley et al. 2005; Hamric and Blackhall 2007), Olson's Hospital Ethical Climate Scale (HEC) (such as Hamric, Borchers, and Epstein 2012), the Hospital Ethical Climate Survey (such as Pauly et al. 2009), the Burnout Scale (such as Meltzer and Huckabay 2004), the Spiritual Well-Being Scale (such as Soleimani et al. 2019), and the Impact of Events Scale (IES; such as De Villers and DeVon 2013) to measure moral distress and identify relationships among variables.

From the many studies conducted in various acute care settings, a number of variables are suggested as possible correlates of moral distress in acute care settings. As examples, moral distress has been found to positively correlate with age and critical care training (Browning 2013); years of experience (Elpern et al. 2005; Hamric et al. 2012); direct patient care (Whitehead et al. 2015; unsafe staffing levels and futile care (Zuzelo 2007); and avoidance (De Villers and DeVon 2013). On the other hand, moral distress has been found to negatively correlate with continuing education, collaboration in patient care conferences, and psychological empowerment (Browning 2013); retention, altered patient care, and spirituality (Cavaliere et al. 2010); age (Corley et al. 2005); spirituality, job satisfaction, self-image, retention, and well-being (Elpern et al. 2005); and ethical climate (Hamric et al. 2012; Whitehead et al. 2015). However, Pauly et al. (2009) report findings from their research based on 374 acute care nurses that reveal no correlation between moral distress and demographics. Corley et al.'s (2001) work also reveals no correlations between moral distress and demographic or professional variables. The fact that the variables that may correlate positively with moral distress have proven difficult to identify speaks to the complexity of the phenomenon of moral distress itself. See Table 4.2 for general categories of contributing factors for moral distress that occur across most studies.

Table 4.1 Recent studies on moral distress in acute care settings

Acute care settings	Studies
Intensive care/Critical care	Allen and Butler (2016), Altaker et al. (2018), Borhani et al. (2017), Browning (2013), Browning and Cruz (2018), Bruce et al. (2015), Burns et al. (2019), Carnevale (2020); Choe et al. (2015), Dacar et al. (2019), De Villers and DeVon (2013), Dodek et al. (2019), Dodeck et al. (2016), Dyo et al. (2016), Elpern et al. (2005), Forozeiya et al. (2019), Fumis et al. (2017), Gibb et al. (2018), Gutierrez (2005), Hamric and Blackhall (2007), Henrich et al. (2017), Hiler et al. (2018), Karagozoglu et al. (2017), Karanikola et al. (2014), Lamiani et al. (2017), Lusignani et al. (2017), Maiden et al. (2011), Mason et al. (2014), O'Connell (2015), McAndrew et al. (2018), Mealer and Moss (2016), Mobley et al. (2007), Puntillo et al. (2001), Rawas (2019), Robinson (2010), Rushton (2016), Saechao et al. (2017), Siffleet et al. (2015), St. Ledger et al. (2013), Van Mol et al. (2015), Wiegand and Funk (2012), Wilson et al. (2013)
Burn unit/Centre	Leggett et al. (2013)
Emergency room	Abdolmaleki et al. (2018), Wolf et al. (2016), Zavotsky and Chan (2016)
Medical/Surgical	Lusignani et al. (2017), Rice et al. (2008)
Mental health/Psychiatry	Deady and McCarthy (2010), Jansen et al. (2019), Lützén et al. (2010)
Oncology (adult)	Ameri et al. (2016), Bohnenkamp et al. (2015), Bruce and Allen (2020), Cohen and Erickson (2006), Lievrouw et al. (2016), Mehlis et al. (2018), Mullin and Bogetz (2018), Orellana-Rios et al. (2018), Pelton et al. (2015), Raines (2000), Vargas Celis and Concha Méndez (2019), Wahlberg et al. (2016)
Operating room	Radzvin (2011)
Palliative/Hospice	Bressler et al. (2017), Lokker et al. (2018), Piers et al. (2012), Taylor-Ford (2013)
Perinatal/Neonatal intensive care	Cavaliere et al. (2010), Dryden-Palmer et al. (2020), Jan Mohamadi et al. (2019), Molloy et al. (2015), Prentice et al. (2020), Prentice et al. (2016)
Pediatrics/Pediatric intensive care	af Sandeberg et al. (2020), Brandon et al. (2014), Dryden-Palmer (2020), Dyo et al. (2016), Garros et al. (2015), Pergert et al. (2018), Thomas et al. (2016), Thomas and McCullough (2016), Trotochaud et al. (2015), Wall et al. (2016)
Renal	Ducharlet et al. (2019)

Table 4.2 Key contributing factors for moral distress[a]

Source	Specifics	Examples
Internal	Individual factors	Individual person-self (e.g., character, values) Interpersonal relationships-others Perception of event-world (e.g., world view, life experiences)
External	Resources	Staffing (e.g., unsafe levels, mix, lack of training/incompetence) Care (e.g., lack of knowledge, lack of information, poor communication)
	Work	Workload (e.g., dangerous) Ethical climate (e.g., poor, toxic)
	Organization	Economic (e.g., cost, efficiency) Regulations (e.g., policies, laws, priorities)

[a]Inspired by Burston and Tuckett (2013)

4.4   Limitations

There are many studies to consider on the topic of moral distress in acute care settings. Limitations generally noted are to be expected, such as a lack of generalizability or a small sample size. However, there are so many common trends across various specialty units and across decades of research that inferences can be safely made. Most studies about moral distress are qualitative or mixed methods. Some studies state that they are strictly quantitative if they only use validated scales, but most still report beyond just the numbers. However, there are no rigorous, randomized controlled trial (RCT) studies on this topic.

When I read about a study, I often ask myself what is missing. I look for information about how many potential participants were invited and how much that number reduced in terms of those who actually completed the study. Did individuals feel that they could decline participation? If they did not, their participation may not have been genuine in terms of their fears about speaking truthfully. Since ethics often involves risk, there are risks to be navigated when/if health care providers choose to participate in research about moral distress and ethical issues in practice. They may be asked to evaluate their workplace satisfaction, their views on their colleagues' competence, or the ethical climate of the organization. Providing feedback on these topics can be risky business and researchers need to recognize the ethics of research on ethics itself. Health care providers may have very real fears of reprisal due to real or perceived issues with autonomy and power dynamics.

Another issue is gender. Most studies on moral distress experiences have a large number of female participants. This is to be expected in nursing. There may be important gender differences with moral distress experiences that deserve further attention. For example, in a national study of Canadian physicians, an e-survey revealed that female practitioners were less satisfied with their careers, had more work-life balance challenges, showed increased frequency and severity of burnout symptoms, and reported greater incivility in their workplace than males (Burns et al. 2019).

One study specifically addresses the gender question with 31 critical care nurses (24 females; 7 males) from an online nursing community (O'Connell 2015). Findings revealed statistically significant higher levels of moral distress reported by females, although overall scores for moral distress were reported to be low in this study. It seems that gender differences need further exploration to determine factors unique to men related to moral distress experiences. Also, lesbian, gay, bisexual, transgender and/or queer (LGBTQ) nurses should have their voices heard. With a history of exclusion and discrimination, LGBTQ nurses (Eliason et al. 2011) likely have moral distress experiences that have not been recognized or explored.

4.5 Conflicting Values

The main sources of moral distress experiences across intensive/critical care settings are noted to be technology, end-of-life care/communication, and the high stress environment (McAndrew et al. 2018). Figure 4.1 depicts an inverted triangle and the roles of proximity to the patient and power of individuals. Nurses have the closest proximity to the suffering of patients in their care in hospital. They have perspective of the details-up close and personal. They also do not hold the same amount of power for final decision making as the most responsible physician (MRP). In the middle are family members and loved ones, who carry more power than they may realize. Many studies reflect on the conflict for health care providers of carrying out family wishes against the needs/wants of the patient. Physicians have the most distance from the day-to-day witnessing of the suffering of individual patients, but perhaps also the most perspective of a 'bigger picture'. This 'incongruence of perspectives' has been described by American researchers using the term 'depth-of-field dissimilarity' where the depth and focus of perspective is anticipated to be specific to the individual (e.g., nurse, family member, physician) (Bressler et al. 2017).

Fig. 4.1 Proximity and power in acute care decision making

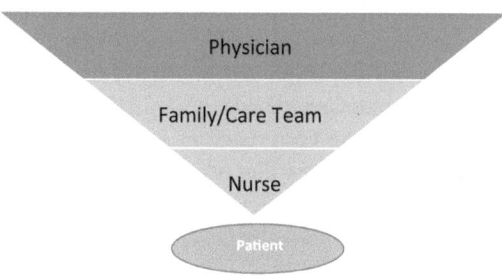

Physician

Family/Care Team

Nurse

Patient

4.6 Death and End-of-Life Decisions

Death is an unforgiving teacher of life lessons in that there is no extra time to right wrongs or have a second chance to do things differently. A key aspect of moral distress experiences among acute care nurses involves palliative and end-of-life care issues, specifically being required to provide futile (potentially inappropriate), non-beneficial, or unnecessary care (Corley 1995; Browning 2013; Elpern et al. 2005; Gutierrez 2005; Hamric and Blackhall 2007; Meltzer and Huckabay 2004; Mobley et al. 2007; Piers et al. 2012; Puntillo et al. 2001; Wilkinson 1987).

Medical assistance in dying (MAiD) in Canada has brought another layer to conversations about death and end-of-life decisions in acute care settings. Between 2015 and 2018 approximately 6,749 individuals chose to have an assisted death which represents just over 1% of the total deaths in Canada (Government of Canada 2019). Not all of these deaths occurred in hospital. A meta-synthesis of the literature on MAiD and nurses' moral experiences cited that nurses in these circumstances are guided by a keen sense of responsibility that is shaped by context and sustained by the moral support of others on their team (Elmore et al. 2016). These findings speak to the relational nature of nursing practice and the process of enactment of moral and ethical decision making.

A Canadian study of 17 nurses' experiences with MAiD revealed a need for ongoing sense making, a wide range of emotions, and three themes that include: (1) role/responsibility of the nurse; (2) personal impact; and, (3) nursing practice (e.g., skills, communication, relational practice) (Beuthin et al. 2018). Some nurses in the study expressed moral distress related to the role of MAiD in their professional practice.

Among health care providers and patients, conscientious objection is often put forward as an issue specific to achieving one person's rights over another, instead of an opportunity to improve communication and foster ethical awareness (Heilman and Trothen 2020). In a concept analysis, Canadian nurse scholar Christina Lamb et al. (2019), describe conscientious objection and conscience as underexplored areas for positive contributions to morality and ethics in nursing practice, theory development, and education. One potential barrier to these opportunities is the close connection of conscientious objection and higher levels of moral distress to established religious beliefs and practices (Davis et al. 2012). Although bringing religious and spiritual aspects into ethics conversations may add complexity and potential discomfort (for some), it is a necessary part of addressing issues from a holistic perspective (Freeman et al. 2020).

4.7 Technology

The range of technological interventions possible in most contemporary acute care settings in developed countries, particularly in critical illness situations at or near the end-of-life, create prime conditions for health care providers to experience moral distress, because time is often limited, pain and suffering are highly likely (and likely high), and the finality of death presents no further opportunities to redeem poor care experiences or incorrect decisions. Families often request that 'everything' be done to contribute to the survival of their loved one. Unfortunately many families do not know what they are asking for when they ask for 'everything'. Diagnostic tests are not cures. Cardiopulmonary resuscitation (CPR) does not redeem someone from a terminal diagnosis. There are limits to modern science that are not apparent on prime time television and Netflix movies.

These are the difficult ethical discussions that occur in the context of '*not can we, but should we*'. Is it 'right' to prolong life at all costs? Different health care professions will have their own perspectives about a patient, with nurses having a 24/7, bedside vantage point. These circumstances do not make a nursing perspective always correct, but it should be worth at least equal consideration.

4.8 Back to Basics

A study examining interpersonal variables and work environment surveyed almost 300 nurses in an American acute care setting (Rathert et al. 2016). Participants identified ethical issues in their practice on a daily to weekly basis (several times a month). Results indicated the need for organizational support for ethics, moral efficacy, and moral distress experiences. But, what specifically can be done and what has been tried and evaluated?

One example of a successful intervention is the use of reflective debriefing guided by a social worker as the facilitator (Browning and Cruz 2018). An American study of 42 ICU nurses on one unit of a hospital, revealed moral distress experiences that were most often related to providing non-beneficial care. Nurses felt empowered when they were able to confront others about lack of truth-telling about a patient's prognosis. This is an example of an interprofessional collaboration that can be supported by organizations who wish to address moral distress among health care providers and utilize resources already available within their staffing complement.

A second example of an intervention is a program of mindfulness, compassion and self-care for a German interdisciplinary palliative care team (Orellana-Rios et al. 2018). The program consisted of a 10-week training led by an experienced meditation teacher. Improvements were found in the following areas: emotional exhaustion, personal accomplishment (with goal attainment), anxiety, stress, rumination, distress, emotional regulation, joy at work, self-care, integration of mindfulness in work routines, interpersonal connections, and team communication.

A third example involves an intervention that was not directly related to addressing moral distress but still had an impact on a number of important areas. In an American study of pediatric care providers at one hospital, a pediatric palliative care program (based on quality of life components) was instituted (Brandon et al. 2014). A cross-sectional survey of health care providers found reductions in the following areas: intensity of providers' distress (specific to work quality of life), frequency of providers' distress (specific to non-beneficial care), impacts of work-related distress on their personal/professional lives, and intention to leave the organization. Health care providers in this study reported more confidence and competence with patient care after this intervention.

> **Recap of Concept: Moral Distress Interventions**
> ✓ Organizational support
> ✓ Reflective debriefing
> ✓ Mindfulness practices
> ✓ Education.

4.9   Conclusion

Moral distress experiences in acute care settings boil down to relationships, communication, and resources. Some scholars have criticized the literature on moral distress for its lack of candid language and downplaying of the seriousness of violating 'bedrock beliefs' and the paucity of interventions to address needs for closure and healing (Morreim 2015). This is an important point that may guide next steps in valuable intervention strategies. A study of almost 700 health care providers working in American ICU settings revealed that the lowest levels of burnout were related to high team work and values alignment between providers and leaders (LeClaire et al. 2019). Further, American nursing ethics scholar Cynda Rushton advocates for attention to moral resilience. Moral resilience attends to: (1) human experience; (2) complexities (e.g., relationships, decisions, obligations); and, (3) challenges (e.g., of conscience, moral distress, confusion) (Rushton 2016). Thus, effective interventions must focus on opportunities to build capacity for moral resilience, personal growth, and professional empowerment among health care providers (Rushton et al. 2016).

References

Abdolmaleki, M., S. Lakdizaji, A. Ghahramanian, A. Allahbakhshian, and M. Behshid. 2018. Relationship between autonomy and moral distress in emergency nurses. *Indian Journal of Medical Ethics* 6:1–5. https://doi.org/10.20529/IJME.2018.076

af Sandeberg, M., C. Bartholdson, and P. Pergert. 2020. Important situations that capture moral distress in paediatric oncology. *BMC Medical Ethics* 21 (6). https://doi.org/10.1186/s12910-020-0447-x

Allen, R., and E. Butler. 2016. Addressing moral distress in critical care nurses: a pilot study. *International Journal of Critical Care and Emergency Medicine* 2: 015.

Altaker, K.W., J. Howie-Esquivel, and J.K. Cataldo. 2018. Relationships among palliative care, ethical climate, empowerment, and moral distress in intensive care unit nurses. *American Journal of Critical Care* 27 (4): 295–302. https://doi.org/10.4037/ajcc2018252.

Ameri, M., Z. Safavibayatneed, and A. Kavousi. 2016. Moral distress of oncology nurses and morally distressing situations in oncology units. *Australian Journal of Advanced Nursing* 33 (3): 6–12.

Anderson, J. 2013. Take action to diminish moral distress in critical care nurses. *American Journal of Critical Care* 22 (4): 286–286. https://doi.org/10.4037/ajcc2013327.

Beuthin, R., A. Bruce, and M. Scaia. 2018. Medical assistance in dying (MAiD): Canadian nurses' experiences. *Nursing Forum* 53 (4): 511–520. https://doi.org/10.1111/nuf.12280.

Bohnenkamp, S., N. Pelton, P. Reed, and C. Rishel. 2015. An inpatient surgical oncology unit's experience with moral distress: Part I. *Oncology Nursing Forum* 42 (3): 308310. https://doi.org/10.1188/15.ONF.308-310.

Borhani, F., A. Abbaszadeh, E. Mohamadi, E. Ghasemi, and M.J. Hoseinabad-Farahani. 2017. Moral sensitivity and moral distress in Iranian critical care nurses. *Nursing Ethics* 24 (4): 474–482. https://doi.org/10.1177/0969733015604700.

Brandon, D., D. Ryan, R. Sloane, and D. Docherty. 2014. Impact of a pediatric quality of life program on providers' moral distress. *MCN American Journal of Maternal/Child Nursing* 39 (3): 187–197. https://doi.org/10.1097/NMC.0000000000000025.

Bressler, T., D.R. Hanna, and E. Smith. 2017. Making sense of moral distress within cultural complexity. *Journal of Hospice & Palliative Nursing* 19 (1): 7–14. https://doi.org/10.1097/NJH.0000000000000308.

Browning, A. 2013. Moral distress and psychological empowerment in critical care nurses caring for adults at end of life. *American Journal of Critical Care* 22: 143–151. https://doi.org/10.4037/ajcc2013437.

Browning, E.D., and J.S. Cruz. 2018. Reflective debriefing: a social work intervention addressing moral distress among ICU nurses. *Journal of Social Work in End-of-Life & Palliative Care* 14 (1): 44–72. https://doi.org/10.1080/15524256.2018.1437588.

Bruce, S.D., and D. Allen. 2020. Moral distress: One unit's recognition and mitigation of this problem. *Clinical Journal of Oncology Nursing* 24 (1): 16–18. https://doi.org/10.1188/20.CJON.16-18.

Bruce, C.R., S.M. Miller, and J.L. Zimmerman. 2015. A qualitative study exploring moral distress in the ICU team: The importance of unit functionality and intrateam dynamics. *Critical Care Medicine* 43: 823–831. https://doi.org/10.1097/CCM.0000000000000822.

Burns, K.E., A. Fox-Robichaud, E. Lorens, and C.M. Martin. 2019. Gender differences in career satisfaction, moral distress, and incivility: a national, cross-sectional survey of Canadian critical care physicians. *Canadian Journal of Anesthesia/Journal canadien d'anesthésie* 66 (5): 503–511. https://doi.org/10.1007/s12630-019-01321-y.

Burston, A.S., and A.G. Tuckett. 2013. Moral distress in nursing: Contributing factors, outcomes and interventions. *Nursing Ethics* 20 (3): 312–324. https://doi.org/10.1177/0969733012462049.

Carnevale, F.A. (2020). Moral distress in the ICU: it's time to do something about it!. *Minerva Anestesiologica*. https://doi.org/10.23736/S0375-9393.19.14021-7

Cavaliere, T.A., B. Daly, D. Dowling, and K. Montgomery. 2010. Moral distress in neonatal intensive care unit RNs. *Advances in Neonatal Care* 10 (3): 145–156. https://doi.org/10.1097/ANC.0b013e3181dd6c48.

Choe, K., Y. Kang, and Y. Park. 2015. Moral distress in critical care nurses: a phenomenological study. *Journal of Advanced Nursing* 71 (7): 1684–1693. https://doi.org/10.1111/jan.12638.

Cohen, J.S., and J.M. Erickson. 2006. Ethical dilemmas and moral distress in oncology nursing practice. *Clinical Journal of Oncology Nursing* 10 (6): 775–782. https://doi.org/10.1188/06.CJON.775-780.

Corley, M.C. 1995. Moral distress of critical care nurses. *American Journal of Critical Care* 4 (4): 280–285. https://doi.org/10.4037/ajcc1995.4.4.280.

Corley, M.C., R.K. Elswick, M. Gorman, and T. Clor. 2001. Development and evaluation of a moral distress scale. *Journal of Advanced Nursing* 33 (2): 250–256. https://doi.org/10.1111/j. 1365-2648.2001.01658.x.

Corley, M.C., P. Minick, R.K. Elswick, and M. Jacobs. 2005. Nurse moral distress and ethical work environment. *Nursing Ethics* 12 (4): 381–390. https://doi.org/10.1191/0969733005ne809oa.

Dacar, S.L., C.L. Covell, and E. Papathanassoglou. 2019. Addressing moral distress in critical care nurses: A systemized literature review of intervention studies. *Connect: The World of Critical Care Nursing* 13(2):71–89. https://doi.org/10.1891/1748-6254.13.2.71

Davis, S., V. Schrader, and M.J. Belcheir. 2012. Influencers of ethical beliefs and the impact on moral distress and conscientious objection. *Nursing Ethics* 19 (6): 738–749. https://doi.org/10. 1177/0969733011423409.

Deady, R., and J. McCarthy. 2010. A study of the situations, features, and coping mechanisms experienced by Irish psychiatric nurses experiencing moral distress. *Perspectives in Psychiatric Care* 46 (3): 209–220. https://doi.org/10.1111/j.1744-6163.2010.00260.x.

De Villers, M.J., and H.A. DeVon. 2013. Moral distress and avoidance behavior in nurses working in critical care and noncritical care units. *Nursing Ethics* 20 (5): 589–603. https://doi.org/10.1177/ 0969733012452882.

Dodek, P.M., M. Norena, N. Ayas, and H. Wong. 2019. Moral distress is associated with general workplace distress in intensive care unit personnel. *Journal of Critical Care* 50: 122–125. https:// doi.org/10.1016/j.jcrc.2018.11.030.

Dodek, P.M., H. Wong, M. Norena, N. Ayas, S.C. Reynolds, S.P. Keenan, S. P., A. Hamric, P. Rodney, M. Stewart, and L. Alden. 2016. Moral distress in intensive care unit professionals is associated with profession, age, and years of experience. *Journal of Critical Care* 31 (1): 178–182. https:// doi.org/10.1016/j.jcrc.2015.10.011

Dryden-Palmer K., G. Moore, C. McNeil, C.P. Larson, G. Tomlinson, N. Roumeliotis, A. Janvier, and C.S. Parshuram. 2020. Moral distress of clinicians in Canadian pediatric and neonatal ICUs. *Pediatric Critical Care Medicine Society of Critical Care Medicine* 21(4): 314–23.

Ducharlet, K., J. Philip, H. Gock, M. Brown, S.L. Gelfand, E.A. Josland, and F. Brennan. 2019. Moral distress in nephrology: Perceived barriers to ethical clinical care. *American Journal of Kidney Diseases*. https://doi.org/10.1053/j.ajkd.2019.09.018.

Dyo, M., P. Kalowes, and J. Devries. 2016. Moral distress and intention to leave: a comparison of adult and paediatric nurses by hospital setting. *Intensive and Critical Care Nursing* 36: 42–48. https://doi.org/10.1016/j.iccn.2016.04.003.

Eliason, M.J., J. DeJoseph, S. Dibble, S. Deevey, S., and P. Chinn. 2011. Lesbian, gay, bisexual, transgender, and queer/questioning nurses' experiences in the workplace. *Journal of Professional Nursing* 27 (4): 237–244. https://doi.org/10.1016/j.profnurs.2011.03.003

Elmore, J., D.K. Wright, and M. Paradis. 2016. Nurses' moral experiences of assisted death: a meta-synthesis of qualitative research. *Nursing Ethics* 25: 1–18. https://doi.org/10.1177/096973 3016679468.

Elpern, E.H., B. Covert, and R. Kleinpell. 2005. Moral distress of staff nurses in a medical intensive care unit. *American Journal of Critical Care* 14: 523–530. https://doi.org/10.4037/ajcc2005.14. 6.523.

Forozeiya, D., B. Vanderspank-Wright, F.F. Bourbonnais, D. Moreau, and D.K. Wright. 2019. Coping with moral distress–The experiences of intensive care nurses: An interpretive descriptive study. *Intensive and Critical Care Nursing* 53: 23–29. https://doi.org/10.1016/j.iccn.2019.03.002.

Freeman, L.A., K.A. Pfaff, L. Kopchek, and J. Liebman. 2020. Investigating palliative care nurse attitudes towards medical assistance in dying: An exploratory cross-sectional study. *Journal of Advanced Nursing* 76 (2): 535–545. https://doi.org/10/1111/jan.14252

Fumis, R.R.L., G.A.J. Amarante, A. de Fátima Nascimento, and J.M.V. Junior. 2017. Moral distress and its contribution to the development of burnout syndrome among critical care providers. *Annals of Intensive Care* 7 (1): 71. https://doi.org/10.1186/s13613-017-0293-2.

Garros, D., W. Austin, and F.A. Carnevale. 2015. Moral distress in pediatric intensive care. *JAMA Pediatrics* 169: 885–886. https://doi.org/10.1001/jamapediatrics.2015.1663.

Gibb, K., K. Then, P. Hruska, and J. Rankin. 2018. The role of the advanced practice nurse in supporting critical care nurses experiencing moral distress. *Canadian Journal of Critical Care Nursing* 29 (2): 55–56.

Government of Canada. 2019. *Fourth interim report on medical assistance in dying in Canada.* Retrieved from https://www.canada.ca/en/health-canada/services/publications/health-system-ser vices/medical-assistance-dying-interim-report-april-2019.html

Gutierrez, K.M. 2005. Critical care nurses' perceptions of and responses to moral distress. *Dimensions of Critical Care Nursing* 24 (5): 229–241.

Hamric, A.B., and L.J. Blackhall. 2007. Nurse-physician perspectives on the care of dying patients in intensive care units: Collaboration, moral distress, and ethical climate. *Critical Care Medicine* 35 (2): 422–429. https://doi.org/10.1097/01.CCM.0000254722.50608.2D.

Hamric, A., C.T. Borchers, and E.G. Epstein. 2012. Development and testing of an instrument to measure moral distress in healthcare professionals. *AJOB Primary Research* 3 (2): 1–9. https://doi.org/10.1080/21507716.2011.652337.

Heilman, M.K.D., and T.J. Trothen. 2020. Conscientious objection and moral distress: A relational ethics case study of MAiD in Canada. *Journal of Medical Ethics* 46 (2): 123–127. https://doi.org/10.1136/medethics-2019-105855.

Henrich, N.J., P.M. Dodek, E. Gladstone, L. Alden, S.P. Keenan, S. Reynolds, and P. Rodney. 2017. Consequences of moral distress in the intensive care unit: a qualitative study. *American Journal of Critical Care* 26 (4): e48–e57. https://doi.org/10.4037/ajcc2017786.

Hiler, C.A., R.L. Hickman Jr., A.P. Reimer, and K. Wilson. 2018. Predictors of moral distress in a US sample of critical care nurses. *American Journal of Critical Care* 27 (1): 59–66. https://doi.org/10.4037/ajcc2018968.

Jan Mohamadi, S., N. Seyd Fatemi, and H. Haghani. 2019. Effectiveness of end-of-life care education on the moral distress of nurses in neonatal intensive care units. *Journal of Pediatric Nursing* 5 (2): 75–82.

Jansen, T.L., M.H. Hem, L.J. Dambolt, and I. Hanssen. 2019. Moral distress in acute psychiatric nursing: Multifaceted dilemmas and demands. *Nursing Ethics.* https://doi.org/10.1177/096697 3019877526.

Lamb, C., M. Evans, Y. Babenko-Mould, C.A. Wong, and K.W. Kirkwood. 2019. Conscience, conscientious objection, and nursing: A concept analysis. *Nursing Ethics* 26 (1): 37–49. https://doi.org/10.1177/0969733017700236.

Lamiani, G., I. Setti, L. Barlascini, E. Vegni, and P. Argentero. 2017. Measuring moral distress among critical care clinicians: Validation and psychometric properties of the Italian moral distress scale-revised. *Critical Care Medicine* 45 (3): 430–437.

LeClaire, M.M., S. Poplau, K. Prasad, C. Audi, R. Freese, and M. Linzer. 2019. Low ICU burnout in a safety net hospital. *Critical Care Explorations* 1 (5): e0014. https://doi.org/10.1097/CCE. 000000000000001.

Leggett, J.M., K. Wasson, J.M. Sinacore, and R.L. Gamelli. 2013. A pilot study examining moral distress in nurses working in one United States burn center. *Journal of Burn Care & Research* 34 (5): 521–528. https://doi.org/10.1097/BCR.0b013e31828c7397.

Lievrouw, A., M.D. Van Belle, and D.D. Benoit. 2016. Coping with moral distress in oncology practice: Nurse and physician strategies. *Oncology Nursing Forum* 43 (4): 505–512. https://doi.org/10.1188/16.ONF.505-512.

Lokker, M.E., S.J. Swart, J.A.C. Rietjens, L. van Zuylen, R.S.G.M. Perez, and A. van der Heide. 2018. Palliative sedation and moral distress: A qualitative study of nurses. *Applied Nursing Research* 40: 157–161. https://doi.org/10.1016/j.apnr.2018.02.002.

Lusignani, M., M.L. Giannì, L.G. Re, and M.L. Buffon. 2017. Moral distress among nurses in medical, surgical and intensive-care units. *Journal of Nursing Management* 25 (6): 477–485. https://doi.org/10.1111/jonm.12431.

Lützen, K., T. Blom, B. Ewalds-Kvist, and S. Winch. 2010. Moral stress, moral climate and moral sensitivity among psychiatric professionals. *Nursing Ethics* 17 (2): 213–224. https://doi.org/10.1177/0969733009351951.

Karagozoglu, S., G. Yildirim, D. Ozden, and Z. Çınar. 2017. Moral distress in Turkish intensive care nurses. *Nursing Ethics* 24 (2): 209–224. https://doi.org/10.1177/0969733015593408.

Karanikola, M.N., J.W. Albarran, E. Drigo, M. Giannakopoulou, M. Kalafati, M. Mpouzika, G.Z. Tsiaousis, and E.D. Papathanassoglou. 2014. Moral distress, autonomy and nurse–physician collaboration among intensive care unit nurses in Italy. *Journal of Nursing Management* 22 (4): 472–484. https://doi.org/10.1111/jonm.12046.

Maiden, J., J.M. Georges, and C.D. Connelly. 2011. Moral distress, compassion fatigue, and perceptions about medication errors in certified critical care nurses. *Dimensions of Critical Care Nursing* 30 (6): 339–345. https://doi.org/10.1097/DCC.0b013e31822fab2a.

Mason, V.M., G. Leslie, K. Clark, P. Lyons, E. Walke, C. Butler, and M. Griffin. 2014. Compassion fatigue, moral distress, and work engagement in surgical intensive care unit trauma nurses: a pilot study. *Dimensions of Critical Care Nursing* 33 (4): 215–225. https://doi.org/10.1097/DCC.000 0000000000056.

McAndrew, N.S., J. Leske, and K. Schroeter. 2018. Moral distress in critical care nursing: The state of the science. *Nursing Ethics* 25 (5): 552–570. https://doi.org/10.1177/0969733016664975.

Mealer, M., and M. Moss. 2016. Moral distress in ICU nurses. *Intensive Care Medicine* 42 (10): 1615–1617. https://doi.org/10.1007/s00134-016-4441-1.

Mehlis, K., E. Bierwirth, K. Laryionava, F.H. Mumm, W. Hiddemann, P. Heußner, and E.C. Winkler. 2018. High prevalence of moral distress reported by oncologists and oncology nurses in end-of-life decision making. *Psycho-Oncology* 27 (12): 2733–2739. https://doi.org/10.1002/pon.4868.

Meltzer, L.S., and L.M. Huckabay. 2004. Critical care nurses' perceptions of futile care and its effect on burnout. *American Journal of Critical Care* 13 (3): 202–208. https://doi.org/10.4037/ajcc2004.13.3.202.

Mobley, M.J., M.Y. Rady, J.L. Verheijde, B. Patel, and J.S. Larsona. 2007. The relationship between moral distress and perception of futile care in the critical care unit. *Intensive and Critical Care Nursing* 23: 256–263. https://doi.org/10.1016/j.iccn.2007.03.011.

Molloy, J., M. Evans, and K. Coughlin. 2015. Moral distress in the resuscitation of extremely premature infants. *Nursing Ethics* 22 (1): 52–63. https://doi.org/10.1177/0969733014523169.

Morreim, H. 2015. Moral distress and prospects for closure. *The American Journal of Bioethics* 15 (1): 38–40. https://doi.org/10.1080/15265161.2014.974771.

Mullin, J., and J. Bogetz. 2018. Point: Moral distress can indicate inappropriate care at end-of-life. *Psycho-Oncology* 27 (6): 1490–1492. https://doi.org/10.1002/pon.4713.

O'Connell, C.B. 2015. Gender and the experience of moral distress in critical care nurses. *Nursing Ethics* 22 (1): 32–42. https://doi.org/10.1177/0969733013513216.

Orellana-Rios, C.L., L. Radbruch, M. Kern, Y.U. Regel, A. Anton, S. Sinclair, and S. Schmidt. 2018. Mindfulness and compassion-oriented practices at work reduce distress and enhance self-care of palliative care teams: A mixed-method evaluation of an "on the job "program. *BMC Palliative Care* 17 (1): 3. https://doi.org/10.1186/s12904-017-0219-7.

Pauly, B., C. Varcoe, J. Storch, and L. Newton. 2009. Registered nurses' perceptions of moral distress and ethical climate. *Nursing Ethics* 16 (5): 561–573. https://doi.org/10.1177/096973300 9106649.

Pelton, N., S. Bohnenkamp, P. Reed, and C. Rishel. 2015. An inpatient surgical oncology unit's experience with moral distress: Part II. *Oncology Nursing Forum* 42 (4): 412–414. https://doi.org/10.1188/15.ONF.412-414.

Pergert, P., C. Bartholdson, K. Blomgren, and M. af Sandeberg. 2018. Moral distress in paediatric oncology: Contributing factors and group differences. *Nursing Ethics* 26 (7–8): 2351–2363. https://doi.org/10.1177/0969733018809806

Piers, R.D., M. Van den Eynde, E. Steeman, P. Vlerick, D.D. Benoit, and N.J. Van Den Noortgate. 2012. End-of-life care of geriatric patient and nurses' moral distress. *Journal of the American Medical Directors Association* 13(1):80.e7–80.e13. https://doi.org/10/1016/j.jamda.2010.12.014

Prentice, T.M., L. Gillam, A. Janvier, and P.G. Davis. 2020. How should neonatal clinicians act in the presence of moral distress? *Archives of Disease in Childhood-Fetal and Neonatal Edition*. https://doi.org/10.1136/archdischild-2019-317895.

Prentice, T., A. Janvier, L. Gillam, and P.G. Davis. 2016. Moral distress within neonatal and paediatric intensive care units: A systematic review. *Archives of Disease in Childhood* 101 (8): 701–708. https://doi.org/10.1136/archdischild-2015-309410.

Puntillo, K.A., P. Benner, T. Drought, B. Drew, N. Stotts, D. Stannard, C. Rushton, C. Scanlon, and C. White. 2001. End-of-life issues in intensive care units: A national random survey of nurses' knowledge and beliefs. *American Journal of Critical Care* 10 (4): 216–229.

Radzvin, L.C. 2011. Moral distress in certified registered nurse anesthetists: Implications for nursing practice. *AANA Journal* 79 (1): 39–45.

Raines, M.L. 2000. Ethical decision making in nurses. Relationships among moral reasoning, coping style, and ethics stress. *JONA'S Healthcare Law, Ethics and Regulation* 2 (1): 29–41.

Rathert, C., D.R. May, and H.S. Chung. 2016. Nurse moral distress: A survey identifying predictors and potential interventions. *International Journal of Nursing Studies* 53: 39–49. https://doi.org/10.1016/j.ijnurstu.2015.10.007.

Rawas, H. 2019. Moral distress in critical care nurses: A qualitative study. *International Journal of Studies in Nursing* 4 (4): 35–41. https://doi.org/10.20849/ijsn.v4i4.659

Rice, E.M., M.Y. Rady, A. Hamrick, J.L. Verheijde, and D.K. Pendergast. 2008. Determinants of moral distress in medical and surgical nurses at an adult acute tertiary care hospital. *Journal of Nursing Management* 16 (3): 360–373. https://doi.org/10.1111/j.1365-2834.2007.00798.x

Robinson, R. 2010. Registered nurses and moral distress. *Dimensions of Critical Care Nursing* 29 (5): 197–202. https://doi.org/10.1097/DCC.0b013e3181e6c344.

Rushton, C.H. 2016. Moral resilience: a capacity for navigating moral distress in critical care. *AACN Advanced Critical Care* 27 (1): 111–119. https://doi.org/10.4037/aacnacc2016275.

Rushton, C.H., M. Caldwell, and M. Kurtz. 2016. CE: Moral distress: a catalyst in building moral resilience. *AJN The American Journal of Nursing* 116 (7): 40–49. https://doi.org/10.1097/01.NAJ.0000484933.40476.5b.

Saechao, N., A. Anderson, and B. Connor. 2017. In our unit: ICU interventions for moral distress and compassion fatigue. *Nursing Critical Care* 12 (1): 5–8. https://doi.org/10.1097/01.CCN.0000503422.13218.67

Siffleet, J., A.M. Williams, P. Rapley, and S. Slatyer. 2015. Delivering best care and maintaining emotional wellbeing in the intensive care unit: The perspective of experienced nurses. *Applied Nursing Research* 28 (4): 305–310. https://doi.org/10.1016/j.apnr.2015.02.008.

Soleimani, M.A., S.P. Sharif, A. Yaghoobzadeh, M.R. Sheikhi, B. Panarello, and M.T.M. Win. 2019. Spiritual well-being and moral distress among Iranian nurses. *Nursing Ethics* 26 (4): 1101–1113. https://doi.org/10.1177/0969733016650993.

St. Ledger, U., A. Begley, J. Reid, L. Prior, D. McAuley, and B. Blackwood. 2013. Moral distress in end-of-life care in the intensive care unit. *Journal of Advanced Nursing* 69 (8): 1869–1880. https://doi.org/10.1111/jan.12053

Taylor-Ford, R.L. 2013. Moral distress in end-of-life care: Promoting ethical standards of executive nursing practice. *Nurse Leader* 11 (3): 51–54. https://doi.org/10.1016/j.mnl.2013.01.005.

Thomas, T.A., and L.B. McCullough. 2016. Resuscitations that never end: Originating from unresolved integrity-related moral distress. *JAMA Pediatrics* 170 (6): 521–522. https://doi.org/10.1001/jamapediatrics.2016.0030.

Thomas, T.A., S. Thammasitboon, D.F. Balmer, K. Roy, and L.B. McCullough. 2016. A qualitative study exploring moral distress among pediatric resuscitation team clinicians: Challenges to professional integrity. *Pediatric Critical Care Medicine* 17 (7): e303–e308. https://doi.org/10.1097/PCC.0000000000000773.

Trotochaud, K., J.R. Coleman, N. Krawiecki, and C. McCracken. 2015. Moral distress in pediatric healthcare providers. *Journal of Pediatric Nursing* 30 (6): 908–914. https://doi.org/10.1016/j.pedn.2015.03.001.

Van Mol, M.M., E.J. Kompanje, D.D. Benoit, J. Bakker, and M.D. Nijkamp. 2015. The prevalence of compassion fatigue and burnout among healthcare professionals in intensive care units: A systematic review. *PLoS ONE* 10 (8): e0136955. https://doi.org/10.1371/journal.pone.0136955.

Vargas Celis, I., and C. Concha Méndez. 2019. Moral distress, sign of ethical issues in the practice of oncology nursing: Literature review. *Aquichan* 19 (1): 11–15. https://doi.org/10.5294/aqui. 2019.19.1.3.

Wahlberg, L., A. Nirenberg, and E. Capezuti. 2016. Distress and coping self-efficacy in inpatient oncology nurses. *Oncology Nursing Forum* 43 (6): 738–746. https://doi.org/10.1188/16.ONF. 738-746.

Wall, S., W.J. Austin, and D. Garros. 2016. Organizational influences on health professionals' experiences of moral distress in PICUs. *HEC Forum* 28 (1): 53–67. https://doi.org/10.1007/s10 730-015-9266-8.

Whitehead P.B., R.K. Herbertson, A.B. Hamric, E.G. Epstein, and J.M. Fisher. 2015. Moral distress among healthcare professionals: report of an institution-wide survey. *Journal of Nursing Scholarship* 47 (2): 117–125. https://doi.org/10.1111kjnu.12115

Wiegand, D.L., and M. Funk. 2012. Consequences of clinical situations that cause critical care nurses to experience moral distress. *Nursing Ethics* 19: 479–487. https://doi.org/10.1177/096973 3011429342.

Wilkinson, J. 1987. Moral distress in nursing practice: Experience and effect. *Nursing Forum* 23 (1): 16–29. https://doi.org/10.1111/j.1744-6198.1987.tb00794.x.

Wilson, M.A., D.M. Goettemoeller, N.A. Bevan, and J.M. McCord. 2013. Moral distress: Levels, coping and preferred interventions in critical care and transitional care nurses. *Journal of Clinical Nursing* 22 (9–10): 1455–1466. https://doi.org/10.1111/jocn.1212.

Wolf, L.A., C. Perhats, A.M. Delao, M.D. Moon, P.R. Clark, and K.E. Zavotsky. 2016. "It's a burden you carry": Describing moral distress in emergency nursing. *Journal of Emergency Nursing* 42 (1): 37–46. https://doi.org/10.1016/j.jen.2015.08.008.

Zavotsky, K.E., and G.K. Chan. 2016. Exploring the relationship among moral distress, coping, and the practice environment in emergency department nurses. *Advanced Emergency Nursing Journal* 38 (2): 133–146. https://doi.org/10.1097/TME.0000000000000100.

Zuzelo, P. 2007. Exploring the moral distress of registered nurses. *Nursing Ethics* 14: 344–359. https://doi.org/10.1177/0969733007075870.

Chapter 5
Community Health Care Contexts

Abstract Moral distress also occurs outside the walls of hospitals, despite persistent beliefs that significant ethical issues must have strict acute care boundaries. However, in community health care settings care providers often have far less control over circumstances and environments. A well-stocked supply cupboard is far from the reach of a visiting nurse who drives, sometimes long distances, to see patients post-operatively for dressing changes and at end-of-life for palliative care. Also explored will be community settings such as public health, primary care clinics, academia, remote First Nation communities, correctional settings, and residential care. These are different stories of moral distress experiences than one might hear in acute care settings, but they retain common ground. Long term and residential care settings appear to be leaders of moral distress research and exploration. This chapter will explore contributions to our understanding of the concept of moral distress from recent academic literature and studies conducted in community care settings.

Keywords Community · Moral distress · Values · Death · Technology

5.1 Introduction

The diverse contexts of community health care settings can present unique ethical challenges that are, at times similar and also often, different from those previously discussed in acute care because of the setting itself. For example a patient's home is borrowed and care interactions may require ongoing negotiation (Stulginski 1993). A borrowed environment changes the dynamics of power and control for health care professionals, patients, and families, and can set the stage for ethical issues that may never surface in an acute care institution. In a hospital, a patient is there with the assumption that they are there to receive care. Should they refuse care and be competent to do so, they can choose to leave and perhaps will be encouraged to do so. The idea that the places and spaces, where care occurs, matter is supported by the nursing practice literature (Andrews 2004; Bender et al. 2007; Liaschenko 2001) and was previously explored in chapter three. As is the mantra for the triad of priorities

© Springer Nature Switzerland AG 2020
K. Jones-Bonofiglio, *Health Care Ethics through the Lens of Moral Distress*,
The International Library of Bioethics 82,
https://doi.org/10.1007/978-3-030-56156-7_5

in real estate, so will be our discussion here: 'location, location, location'. Location matters because context matters.

To begin on a positive note, community care settings can be uniquely therapeutic places of health and healing. Caring for patients in their own established social and cultural milieu can be helpful to maintain as many personally valued roles and responsibilities as possible and as such, contribute to healing and wellbeing. However, sometimes the environment itself can bring its own barriers. For the purposes of this chapter, a number of areas of 'community care' will be explored in the context of moral distress experiences among nurses, including home care, public health, primary care clinics, post secondary academic environments (e.g., colleges and universities), military and veterans services, correctional settings, and long-term care/residential care. Other community care environments and diverse health care professions will be noted.

Marginalization and vulnerability can be found across all health care sectors, but perhaps it is in community settings where individuals' life circumstances are most visible, tangible, and undeniable. The impact of the social determinants of health cannot be overlooked. In community health care settings ethical issues can be complex, requiring increased collaboration among health care providers, ongoing role clarification, and attention to family dynamics (MacPhail 2001) at much deeper levels than acute care providers may ever experience. Key everyday ethical issues for nurses in community care settings include lack of time, increased client autonomy, the need for client support systems, variable family support, health care team communication issues, and lack of resources (Burger et al. 1992). These examples of ethical issues may seem similar to ones discussed for acute care, however often the underlying factors are specific to the community context in which they occur.

Further, my own master's thesis was a study among professionals (i.e., nurses and social workers) in community health care settings in northern Ontario, Canada. Self-reports revealed findings of moderate to high levels of frequency and intensity of moral distress experiences in everyday practice among these health care professionals (Jessiman 2008). Participants also expressed a moderate to high level of confidence in their own perceived ability to recognize, address, and resolve ethical issues in their practice. This research supports the notion that while health care providers in community health care settings report that they generally feel confident in their capacity to managing everyday ethical issues, there remains a very significant underlying issue of moral distress experiences. Of note in this research, is that I conducted the work using a paper survey that was mailed back to me. When I received the surveys back, the participants had used any available blank spaces on the surveys to write out their experiences of ethical issues and moral distress. I had unwisely dedicated only a small space at the end of the survey for 'any additional comments'. The participants' stories flowed over into every last empty space that was available; that was my 'ah-ha' moment that there is much more work to be done in community to support ethics in practice.

In addition, there are a number of barriers to ethical nursing practice that have been identified in the academic literature which may contribute to moral distress experiences in community settings. Bowman (1995) asserts that a nurse's response to ethical issues is influenced by constraints, expectations, protocols, and perceived powerlessness. Roles and responsibilities for nurses, especially in community care settings, may compete and conflict with patient needs, patient choice and autonomy, family or caregiver requests and expectations, organizational/agency demands, requirements of the job (i.e., working extra hours unpaid), and individual nurse's personal and professional needs (Schoot et al. 2006). These findings also resonate with Meaney's (2002) concept of multiple masters, where the needs of a variety of stakeholders result in decision making that is complex and context-laden.

5.2 Moral Distress Experiences in Community

At this time, there is only a small amount of empirical research and academic literature on moral distress for nurses working and caring for patients and families in community health care settings. Moral distress has, however, been explored in community settings in a handful of studies which will be discussed here. Ethics in community health care practice requires a different ethical approach that considers interdependence, inclusion, diversity, advocacy, social justice, participation, and empowerment (Racher 2007). Reviewing the recent literature on moral distress in community care settings is foundational for two reasons: (1) to identify the spectrum of areas in community care in which researchers have identified moral distress experiences; and, (2) to highlight potential contemporary variables that appear to correlate with nurses' moral distress experiences, as well as strategies and recommendations. Many areas of nursing practice, within community settings, are linked to studies on moral distress (see Table 5.1). Clearly, the academic literature supports the presence of the phenomenon of moral distress in community care settings.

5.2.1 Home Care

As an example of a study from the home care services sector, Canadian researcher Kevin Brazil et al. (2010) examined experiences of moral distress among care providers in home-based palliative care settings in southern Ontario, Canada. Interviews were conducted with 18 care providers from five home-visiting organizations. The care providers included personal support workers (PSWs, also known as unregulated care providers [UCPs]); rehabilitation therapists; social workers; and, nurses. Three themes about moral distress experiences emerged from analysis of the data:

(1) the role of the informal caregiver (e.g., caregiver burden, neglect or abuse, or competency to provide care);
(2) challenging issues in practice (e.g., respect for patient decision making, inappropriate treatment, pain and suffering, unanticipated death, and communicating information about death and dying); and,

Table 5.1 Moral distress in community care settings

Community settings	Studies author (year)	Contributing factors
Community learning disabilities care	Holloway (2004)	• Policies, isolation, and time
Continuing care (in facilities and at home locations)	Ritchie, O'Rourke, and Stahlke (2018)	• Policies, palliative care, perceptions (of nurse practitioner role/scope of practice), physician power struggles, and complex patient needs
Pharmacy practice	Astbury, Gallagher, and O'Neill (2015)	• Highly regulated profession, strict legal framework, professional code of conduct, and value-based profession
Residential child care	McMillan (2020)	• Contractual regulations and the commodification of care
Tele-health care	Rutenberg and Oberle (2008)	• Ambiguous work environment, time of day, weather, distance from care/transportation, and policies

(3) the service delivery system (e.g., lack of resources, incompetence among colleagues, and lack of timely client information delivery to care providers).

This study reveals the unique relationships between informal caregivers (e.g., family members) and professional health care providers in home-based care environments as a key source of moral distress.

In home care of clients with chronic illness, 10 Dutch nurses were interviewed about strategies to balance competing roles and responsibilities (Schoot et al. 2006). Nurses noted that they used detaching, pleasing, dialoging, and directing as the main options to cope with moral conflicts. Recommendations from this study to support ethical, quality nursing practice include taking a relational approach with the client, self-assertiveness, professional autonomy, critical ethical reflection, and organizational support. These strategies also support nurses' empowerment.

In an American study of 15 home care nurses, psychological empowerment was described as being held within the nurse-patient-caregiver relationship and involves confidence, competence, independence, trust, comfort, collaboration, meaning, and choice (Williamson 2007). Participants in this study described empowerment as having the knowledge and skills to respond to and actively partner with others to meet complex care needs.

In home care and from a psychiatric care perspective, Swedish researchers conducted focus groups with nurses on ethical decision making and patient autonomy in home care settings (Magnusson and Lützén 1999). Participants described uncertainty and conflicting responsibilities in their professional nursing role. Key themes included: intruding into someone's home; fluctuating boundaries; respecting privacy; and, mutual vulnerability. Although this study is over 20 years old the themes appear to remain relevant today. Community health care providers continue to require knowledge, skills, and attitudes that will help them to address diverse ethical issues in everyday practice.

5.2.2 Public Health

In contrast to an individual approach to ethical decision making, one must also consider the tensions for ethical decisions that involve larger groups of individuals, such as communities. According to the Public Health Agency of Canada (2008), the seven core competencies for public health professionals are:

- Public health sciences;
- Assessments and analysis;
- Policy/program planning, implementation, and evaluation;
- Partnerships, collaboration, and advocacy;
- Diversity and inclusiveness;
- Communication; and,
- Leadership.

Broadly speaking then, public health care is focused on promoting health and on preventing illness and injury of populations and communities (sometimes from a global perspective) with the goal of achieving the common public good (Kenny et al. 2010). The moral considerations involve a balancing of benefits over harms for a population versus for any one individual (Weijer et al. 2013). A public health focus, therefore, leads to a different discourse from the above discussions on an individual's rights of autonomous choice and personal decision making. As such, the moral foundation for contemporary public health practice is social justice, with special attention paid to addressing inequalities and correcting injustices for the most vulnerable members of society (Kenny et al. 2010; Powers and Faden 2006). Population and public health ethics in Canada is guided by organizations, such as the Canadian Institutes of Health Research-Institute of Population and Public Health (CIHR-IPPH), the Public Health Agency of Canada (PHAC), and the National Collaborating Centre for Healthy Public Policy (NCCHPP), that are working collaboratively to advance and support work in this area (Viehbeck et al. 2011).

Let's begin with a public health nursing example with an international perspective. Using a cross-sectional, survey study design, public health nurses (PHNs) from public health units across Japan were asked about ethical issues in their practice (Asahara et al. 2012). Researchers found that ethical issues were often identified when nurses' views were different from client and family views. However, in consideration of Japanese culture, the division between patient and family views is much less defined than in western culture. In fact, Japanese families may even routinely make decisions for the patient without consultation. Further, ethical issues were identified between nurses and administration and nurses and other organizations. This study recommends ongoing ethics education for nurses and means to support ethics concerns in everyday practice.

A Canadian study of 22 public health nurses, explored ethics in practice and found the need for a relational response that attempts to maximize good outcomes (Oberle and Tenove 2000). Five themes emerged from the analyzed data and these included: (1) relationships with other professionals; (2) systems; (3) character of relationships; (4) respect; and, (5) putting self at risk. This study is 20 years old and the findings still echo across community settings and across health care sectors.

5.2.3 Primary Care Clinics

A study of 71 American nurse practitioners (NPs) explored moral distress and ethical issues in primary care practice through the use of self-report on an original survey (Laabs 2005). Ethical dilemmas most often arose due to patients' refusal of recommended treatment (described as patient autonomy versus NP beneficence). Moral distress experiences were described by NPs as situations of frustration and powerlessness. NPs considered changing their employment location and/or leaving advanced nursing practice.

What could be done? In a study of physicians in primary care a brief mindfulness interventions was tried and evaluated (Fortney et al. 2013), showing improvements in symptoms of burnout, better mental health (e.g., anxiety, depression), and reduced stress levels. This type of approach does nothing to address the underlying issues but may still be useful to health care providers' overall health and wellbeing.

5.2.4 Academia

Faculty members in nursing programs, often graduate degree prepared, face a very different reality than that of their predecessors in today's college and university settings. Academic settings are high stakes environments (Tippitt et al. 2009), where nursing professors face ethical issues of incivility (Beasley 2010), academic dishonesty, and grade inflation that may lead to job dissatisfaction among professors and negative impacts to student learning (Ganske 2010).

A study by Canadian nurse researchers Singh et al. (2014), highlights organizational culture, psychological empowerment, and structural empowerment as the keys to good, ongoing support and successful integration of nurse scholars into this unique setting. In this study, a healthy work environment was designated by a manageable workload, available mentorship, both horizontal (from colleagues) and vertical (from senior leadership/administration), and collegiality (safe interactions). However, as in many environments for health care providers, a responsive work environment is not always readily available.

In response to ethics needs in academic settings, two American nurse scholars revised a code of ethics for nurse educators (Rosenkoetter and Milstead 2010) that was original published almost thirty years prior (Rosenkoetter 1983). These scholars noted the need for the revision in response to changes in the practice environment related to new technologies (in the classroom and in health care), society's expectations, and advances in nursing science for a global community. The revised code offers 17 ethical responsibilities for nurse educators and claims to meet the needs of both educators and students. However, the ethical responsibilities of students themselves are not noted in either the original code or the revised one.

In Canada, nursing practice is regulated provincially and therefore, there are subtle differences between each province and territory. Although specific ethical guidelines for student nurses do not exist, in the Canadian Nurses Association's (2008) *Code of Ethics for Registered Nurses* it is stated that "Nursing students are expected to meet the standard of care for their level of nursing" (p. 50) and, if unable to do so, are to notify their clinical instructor and the unit's supervisor. As ethical care is expected across all nursing standards, students need to be educated regarding ethical practice and of this requirement (which, unfortunately does not always occur).

While post-secondary institutions generally require their students to adhere to a student code of conduct and follow regulations for academic integrity, health care professions students are not usually regulated under different ethical standards than, say, a history major or a philosophy student. This can make it difficult to uphold a high level of ethics in clinical nursing practice when the academic baseline (written rules) may not have a clear crossover into clinical practice. Further, these academic regulations may be appealed upon disciplinary action and judged by a non-health care professions academic in terms of appropriateness of the sanctions that have been applied.

Nursing students have also been studied in terms of their experiences of ethics in practice and responses of moral distress. In a Brazilian study, the Moral Distress Scale for Nursing Students was given to 499 Brazilian nursing students and revealed higher levels of moral distress among students in their most senior years in the nursing program (Bordignon et al. 2019). Further, the students described how they used strategies of resistance and demonstrated moral empowerment when faced with morally distressing experiences (Bordignon et al. 2018).

In a Canadian study, seven nursing students described their moral distress experiences in a 13-week clinical placement on a mental health in-patient unit (Wojtowicz et al. 2014). The students reported first hand experiences of a lack of relational practice between nurses and patients, physicians' hierarchy of power, and the inability of their clinical instructor to effect change. As a result of these experiences, some students noted that they would not want a nursing career in mental health. Further, nursing instructors have witnessed their students' moral distress while in clinical placement and have expressed their own frustration, powerlessness, and sense of responsibility to support the students during these experiences (Wojtowicz and Hagen 2014).

5.2.5 Military and Veterans Care

Almost two decades ago, an American study spearheaded the development of a model of moral distress in military nursing based on Jameton's conceptual work (1984, 1993) via interviews with 13 Nurse Corps officers (Fry et al. 2002). From the study, the researchers describe a variety of settings for military nursing practice that all involve serious risks for physical, emotional, psychological, and spiritual stressors related to working in war zones. Nurses shared stories of their commitments to others and how not being able to carry out the right action was a deep violation of their core values as people and as military nurses. It is interesting to note that participants saw themselves as not only part of nursing culture, but also part of military culture; navigating two world views.

Military culture is a unique world of its own. American military nurse practitioner Cynthia Kuehner writes of the need to more fully understand military culture in order to provide better health care to veterans (Kuehner 2013). The author describes military ethos, such as: honour, courage and commitment (Navy core values); integrity, service, excellence (Air Force core values); and, timeless faithfulness (Marine core values). Perhaps the need to further understand fundamental core values in the context of military nursing is crucial to providing effective approaches to address moral distress in military settings.

5.2.6 Correctional Settings

An Italian study undertook a preliminary validation of the Moral Distress Scale for Correctional Nurses with 238 nurse participants in various correctional settings (Lazzari et al. 2020). They found moderate levels of moral distress, which were most closely linked with longer nursing experience, (perceived) incompetence of colleagues, intention to leave, and low staffing levels.

In another Italian study of nurses in correctional institutions, a mixed methods approach was taken to explore ethical issues in practice as well as job satisfaction (Carnevale et al. 2018). Nurses' reported ethical issues related to relational, structural, and organizational factors. The qualitative findings revealed five main themes that relate well to issues seen generally across moral distress experiences. These include: (1) being able to meet patients' needs; (2) negotiating boundaries of care; (3) job satisfaction; (4) barriers to good care; and, (5) prioritizing task management.

5.2.7 Remote First Nation Communities

There is perhaps no other type of Canadian community nursing that faces as much difficulty with recruitment and retention as agencies attempting to staff services for remote First Nation communities. Therefore with the well documented link between moral distress and intent to leave, addressing moral distress among health care providers in these unique settings is very important. A study of three Ojibway communities in northern Ontario, Canada, explored the serious consequences to continuity of care due to high turnover rates and the use of short-term relief nurses (Minore et al. 2005). In these communities, which are often accessible only by air, nurses are the primary health care providers. Two consequences of the limitations nurses faced were a focus on acute care only and patient disengagement. I suggest that these are consequences for the nurses, as well as the patients in the community, in terms of the nurses' job satisfaction and feeling that their professional responsibilities are not being fulfilled. Further, such circumstances leave little room for nurses to establish therapeutic relationships, learn about language and culture, and to earn much needed trust with community members.

5.2.8 Long-Term/Residential Care

In a study of seven long term care (LTC; or residential care) facilities in Ontario, Canada, the experiences of 30 externally hired (unregulated) companions were explored (Brassolotto et al. 2017). Participants in this study were mostly women, racialized, and/or not born in Canada. Since these private companion workers were hired to care for just one individual, one might suspect that the risk for moral distress would be quite low with factors such as time and workload being minimized. In fact, researchers found that the same structural barriers within the organization that potentiates moral distress among internal staff, were factors in moral distress experiences for these workers. Further, paid companions witnessed the staff's struggles to care properly for residents and experienced moral distress because they wanted to assist (and either eventually did [breaking protocol] or didn't [outside scope of practice]).

Across the pond in England, interviews of 16 care staff (i.e., nurse managers, nurses, health care aids [HCAs]) across four LTC homes explored experiences of caring for residents at end-of-life (Young et al. 2017). Four key themes emerged: relating, caring, communicating, and advocating. These strategies were found to either influence the ability to take appropriate action or lead to a sense of powerlessness. If the right action could not be sought, care staff often described that a resident's death was not a 'good death'. A foundational source of moral distress experiences was described as incongruent values that could occur between staff, residents, relatives, and physicians.

Caregiving of older adults with dementia in LTC appears to be a type of care that is linked to high risk for moral distress. In another Canadian study, researchers in Alberta used the Moral Distress in Dementia Care Survey with 389 care providers (i.e., registered nurses [RNs], licensed practical nurses [LPNs], and HCAs) at 23 different long term care sites (Pijl-Zieber et al. 2018). Findings showed a frequency of moral distress experiences that ranged from daily to weekly occurrences with most reporting a moderate intensity. They noted that both frequency and intensity of moral distress experiences positively correlated with proximity to residents' care. Moral distress experiences were also linked with intentions to quit.

Further, a second study in Alberta, Canada, explored moral distress experiences of 18 care staff (participants included RNs, LPNs, and HCAs) across six LTC facilities and three home care agencies (Spenceley et al. 2017). Staff disclosed moral distress experiences related to conflicting care expectations (e.g., inappropriate, futile, inadequate, delayed), wrong model of care/level of care to meet residents' needs, and working in a culture that values tasks over touch. Issues that led to feelings of powerlessness were linked to family requests, physician's orders, responses of leadership, and actions by other care providers. Participants described the strategy of 'staying silent' and using deception, and at times felt 'bullied' to do so. In a further publication on this study, the need for supportive and responsive leadership to manage and mitigate moral distress experiences among LTC staff was noted (Spenceley et al. 2019).

In an attempt to accurately measure moral distress in the context of aged care, Australian researchers amended the Moral Distress Scale-Revised (MDS-R) and had 106 care provider participants complete the survey (Burston et al. 2017). Mean scores showed low levels of frequency, but when moral distress did occur, they showed moderate intensity. Moral distress experiences were associated with low staff competency issues, poor care, and poor communication.

From a German perspective, interviews with 21 nurses across five nursing homes found moderate to severe intensity of moral distress experiences (Kada and Lesnik 2019). Key categories linked to moral distress included: end-of-life care, resident's behaviour management, poor care, unnecessary hospital transfers, and non-agreement with relatives or physicians. Moral uncertainty was noted, especially when the resident's wishes were unclear or a physician was unavailable.

In a Swedish study of nurses working in residential care for older adults, eight registered nurses shared their paradoxical experiences (Karlsson et al. 2009). They described feeling appreciated and undervalued, and having clear roles and endless responsibilities. Using the term 'lonely fixers', they described expectations that they would routinely participate in complex problem solving and provide specialized care without necessary team expertise. This study suggests the need for communication of clear professional boundaries (e.g., scope of practice), manager's support and supervision, and ongoing, mandatory professional development.

Finally, a Canadian study explored 15 nurses' experiences of 'initial moral distress' in LTC (Edwards et al. 2013). Three themes emerged, namely: context, team, and outside direction. Assistance for moral distress experiences was often sought from supportive team members, managers, and educational opportunities. They noted that the 'in-between' position of nurses in the health care system is a contributing factor to moral distress experiences.

Canadian nursing ethics scholars Colleen Varcoe, Gweneth Doane, Bernadette Pauly, Paddy Rodney, Janet Storch, Rosalie Starzomski (and colleagues) have written extensively on moral distress and also about the concept of nurses working the 'in-betweens' (Varcoe et al. 2004). Their study explored nurses' perspectives on ethics in practice and found that nurses define ethics as both 'being' (a way of being) and

Fig. 5.1 Nurse in the middle

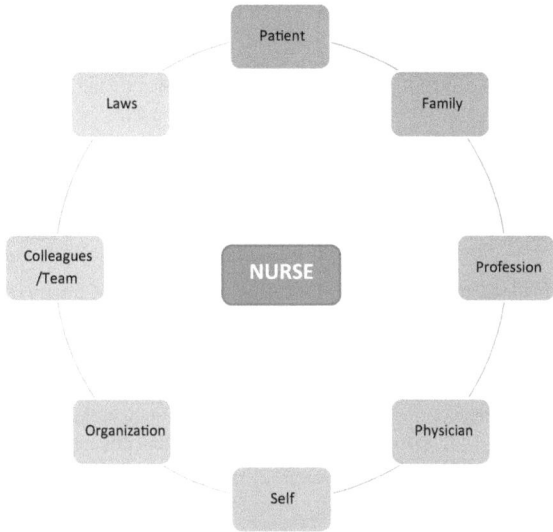

'doing' (a process of enactment). The nurses described struggling to work the 'in-betweens' of their values and organizational values, other's values, and competing values/interests. American nursing ethics scholar, Ann Hamric, writes of this and uses the term 'being in the middle' to describe nurses' positionality of having multiple fidelity and fiduciary duties (Hamric 2001). See Fig. 5.1 for further details.

5.3   Common Denominators

Within this field of scholarship, a number of variables are suggested as possible correlates of moral distress in community care settings (see Table 5.2). Some root causes of moral distress in acute care settings include: real/perceived powerlessness, lack of knowledge, inadequate staffing, lack of support, incompetence, futile treatment/aggressive care, inadequate consent, and lack of truth telling (Hamric 2012). These root causes from acute care are compared as they occur in the context of community care settings using examples cited in this chapter. See Table 5.2.

5.4   Limitations

Most studies noted in this chapter are either qualitative or mixed methods studies. As such, they carry the usual limitations that should not need to be apologized for (namely that they don't contain the same qualities as quantitative studies) but generally are, such as small sample size, lack of randomization, lack of control and

Table 5.2 Correlates of moral distress in community care settings

Acute care contexts	Community care contexts
Real/perceived powerlessness	Policies that do not support ethics in practice. Physician power struggles. Nurses' conflicting responsibilities. Patient wants/needs versus family wishes. Patient refusal of treatment recommendations. Feeling bullied.
Lack of knowledge	Informal/family caregivers. Colleagues. Uncertainty about role/responsibilities. Wrong model/level of care. Culture that values tasks over touch. Questionable physician's orders.
Inadequate staffing	Isolation. Lack of time. Heavy workloads. Low staffing levels. Delayed care. Poor quality care.
Lack of support	Ambiguous care environment. Isolation. Distance from care/transportation. No time to talk with colleagues.
Incompetence	Complex patient needs. Unnecessary hospital transfers. Incompetent colleagues.
Futile treatment/aggressive care	Inadequate palliative care/end-of-life care. Inappropriate treatments. Focus on acute health needs only. Lack of resources.
Inadequate consent	Unclear patient wishes. Following through on family wishes versus patient's wants/needs.
Lack of truth telling	Communication (e.g., quality, timeliness). Staying silent. Using deception.

experimental groups, unequal group sizes, etc. Most studies were conducted in major urban centres and this has implications for application in other locations (e.g., rural, remote).

It is important to note that in these studies, participants self-selected to take part. As such, we must give consideration to those whose voices go unheard in this research. Are the care providers who are not participating the ones who already know how to cope with moral distress? Or are these care providers the ones that have been most deeply impacted by moral distress, to the point that they would not even consider participating in a study? Sometimes the participants that researchers most need to access and understand are the ones that do not volunteer their time.

In comparing and contrasting these studies, consideration must be given for various political, historical, economic and cultural factors that may not be obvious across different countries and across decades of time. However, most studies are a snapshot of one point in time, of a small group of health care providers, in a specific location and context. Also, studies in community often mix different groups of participants, which is something not usually seen in studies conducted in acute care settings. Community settings often have more unregulated health care staff and the term 'nurse' may be colloquially extended.

The concept of moral distress is, in itself, difficult to study as it is an intangible concept and most studies reveal subjective accounts of personal experiences.

However, the commonalities between the handful of studies noted here and the diversity of many types of community settings is an interesting phenomenon. There are clearly areas of common ground that have remained consistent across time and in different contexts.

5.5 Conflicting Values

The benefits and burdens of relationships and spatial dynamics in community care settings shine through in many of these studies. In community there is often more time to connect with patients and families (over time, not necessarily per visit) as care may occur across months or years and in the realm of a person's 'home' environment. This closeness or proximity to patients has been referred to as being a 'proximal nurse' versus being a 'distal nurse' (Malone 2003). Physical and emotional proximity has been proposed as also having moral and narrative proximity. The argument is that acute care nurses are increasingly forced to practice distally with a singular focus on procedural efficiency. It may be that nurses in community have the potential benefit of being more proximally located in their practice. However, a closer connection may also set the stage for increased frequency and/or intensity of moral distress experiences and moral ambiguity (e.g. not able to see the forest for the trees) (Peter and Liaschenko 2004).

> **Recap of Concept: Moral Distress in Community Health Settings**
> ✓ Relationships
> ✓ Power
> ✓ Policies
> ✓ Conflicting responsibilities
> ✓ Communication

5.6 Death and End-of-Life Decisions

Death offers powerful teachings about life (Singer 2007). A strong theme across many studies of moral distress experiences is unease related to palliative and end-of-life care. In acute care settings, futile or non-beneficial treatments, in the context of too much care, are often a source of moral distress. In community, the opposite is often found, it terms of having too little to offer (e.g., services, resources, expertise).

Palliative care in community may be best delivered using a model of short-term integrated palliative and supportive care (SIPS) (Bone et al. 2016). This model aligns benefits, timing and processes, and collaboration between specialists and generalists. Further, it enhances quality care through opportunities for discussion, holistic assessment, pain and symptom management, and caregiver support.

In Canada, medical assistance in dying (MAiD) in the community setting is new territory for many health care providers. Access to physicians who will support this legal option in various community settings is varied across the country and limited in northern, rural, and remote settings. A patient's or family's idyllic version of a 'peaceful death at home' actually requires significant planning and collaboration (Ladd et al. 2000). Appropriate and adequate professional care and support with knowledgeable health care providers, functional policies and procedures, and good communication is still a work in progress in many cases.

5.7 Technology

Of note, technology was not directly mentioned in studies reviewed for this chapter. Technology always plays a role in health care and nursing practice, however it is often overlooked as a source of ethical conflict and moral distress experiences. Shifting boundaries between digital worlds and face-to-face health care environments are changing the ways nurses practice and expanding opportunities for patients to connect to services and providers. This may be an unanticipated source of moral uncertainty in community care settings.

Health care providers need a keen sense of ethics in practice to navigate dual relationships, boundary crossings, and boundary violations (Powers 2003). Boundary crossings can be defined as actions and behaviours that are still deemed to be therapeutic but enter the 'grey zone' between personal and professional. Boundary violations are actions and behaviours that benefit the provider more than the patient and are related more directly to breaches of standards of professional ethics.

5.8 Back to Basics

In an Australian study of nurses and their work in aged care facilities, nine RNs were interviewed to reveal eight common themes about resilience in their nursing practice (Cameron and Brownie 2010). Nurses spoke of the protective value of being able to establish and maintain meaningful relationships with residents in their care. Their preferred go-to for support was their colleagues for the purposes of debriefing, validation, or a much needed laugh (Fig. 5.2).

In a Norwegian study of community nurses' job engagement, researchers found that nurses who were able to thrive despite adversity were the ones who made a habit of regularly practicing active self-reflection and purposive introspection (Vinje and Mittelmark 2008). This allowed them to find meaning and seek positive adaptations to their work. These nurses expressed joy, enthusiasm, and high commitment. Out

Fig. 5.2 Transforming
moral distress experiences in
community care settings

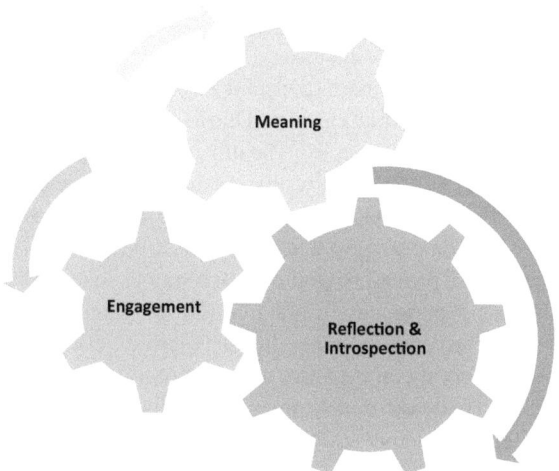

of the 11 nurses identified as 'superstars', nine revealed that they had experienced symptoms of burnout at some point in their career. See Fig. 5.1 for key aspects of nurses' resilience (Table 5.3).

Table 5.3 Resilience[a]

Roots	Growth	Benefits (that nourish the roots)
Experience	Competence Confidence	Time management Crisis management Prioritizing tasks/staff Flexibility
Satisfaction	Clinical skills and knowledge	Holistic care Compassion, empathy
Positive attitude	Feel that you are making a difference	Self-reflection Faith
Relationships	Share with others Mentorship Cohesive teams	Help/support Debriefing Validating Camaraderie
Insight	Recognize stressors	Put strategies in place for stress management
Self-care	Maximize work-life balance	Maintain social supports/interests Exercise Rest/relaxation

[a]Inspired by Cameron and Brownie (2010), further content added.

5.9 Conclusion

One factor that remains constant across various community health care settings is the strong bonds of relationships. Due to a different kind of proximity, provision of care in community is distinct from health care delivery in traditional acute care settings (i.e., hospitals). In community, relationships with patients and families can be closer, occur over longer periods of time, and have meaning beyond just a patient-care provider context. Some suggest that building a stronger moral community among and between health care providers would be one way to sustain moral integrity and address moral distress experiences in practice (Hardingham 2004). Others suggest that it is the health care environments themselves (i.e., climate, culture, and relationships) that are the moral communities and thus require attention and capacity building (Austin 2007). Next, we will explore moral distress from the foundations of personal and professional identities and potential impacts on team and organizational factors.

References

Andrews, G.J. 2004. (Re) thinking the dynamics between healthcare and place: Therapeutic geographies in treatment and care practices. *Area* 36 (3): 307–318. https://doi.org/10.1111/j.0004-0894.2004.00228.x.

Asahara, K., M. Kobayashi, W. Ono, J. Omori, H. Todome, E. Konishi, and T. Miyazaki. 2012. Ethical issues in practice: A survey of public health nurses in Japan. *Public Health Nursing* 29 (3): 266–275. https://doi.org/10.1111/j.1525-1446.2011.00990.x.

Astbury, J.L., C.T. Gallagher, and R.C. O'Neill. 2015. The issue of moral distress in community pharmacy practice: Background and research agenda. *International Journal of Pharmacy Practice* 23 (361–366): 457. https://doi.org/10.1111/ijpp.12174.

Austin, W. 2007. The ethics of everyday practice: Healthcare environments as moral communities. *Advances in Nursing Science* 30 (1): 81–88.

Beasley, S. 2010. Nurse educators: Stop the bullying! *The South Carolina Nurse* 10 (3): 6.

Bender, A., L. Clune, and S. Guruge. 2007. Considering place in community health nursing. *Canadian Journal of Nursing Research Archive* 39 (3): 20–34.

Bone, A.E., M. Morgan, M. Maddocks, K.E. Sleeman, J. Wright, S. Taherzadeh, C. Ellis-Smith, I.J. Higginson, and C.J. Evans. 2016. Developing a model of short-term integrated palliative and supportive care for frail older people in community settings: Perspectives of older people, carers and other key stakeholders. *Age and Ageing* 45 (6): 863–873. https://doi.org/10.1093/ageing/afw124.

Bordignon, S.S., V.L. Lunardi, E.L.D. Barlem, G.D.L. Dalmolin, R.S. da Silveira, F.R.S. Ramos, and J.G.T. Barlem. 2019. Moral distress in undergraduate nursing students. *Nursing Ethics* 26 (7–8): 325–2339. https://doi.org/10.1177/0969733018814902.

Bordignon, S.S., V.L. Lunardi, E.L. Barlem, R.S.D. Silveira, F.R. Ramos, G.D.L. Dalmolin, and J.G.T. Barlem. 2018. Nursing students facing moral distress: Strategies of resistance. *Revista Brasileira de Enfermagem* 71: 1663–1670. https://doi.org/10.1590/0034-7167-2017-0072.

Bowman, A. 1995. Teaching ethics: Telling stories. *Nurse Education Today* 15: 33–38. https://doi.org/10.1016/S0260-6917(95)80076-X.

Brassolotto, J., T. Daly, P. Armstrong, and V. Naidoo. 2017. Experiences of moral distress by privately hired companions in Ontario's long-term care facilities. *Quality in Ageing and Older Adults* 18 (1): 58–68. https://doi.org/10.1108/QAOA-12-2015-0054.

Brazil, K., S. Kassalainen, J. Ploeg, and D. Marshall. 2010. Moral distress experienced by health care professionals who provide home-based palliative care. *Social Science & Medicine* 71 (9): 1687–1691. https://doi.org/10.1016/j.socscimed.2010.07.032.

Burger, A.M., J.A. Erlen, and L. Tesone. 1992. Factors influencing ethical decision making in the home setting. *Home Healthcare Nurse* 10 (2): 16–20. https://doi.org/10.1097/00004045-199203 000-00004.

Burston, A., R. Eley, D. Parker, and A. Tuckett. 2017. Validation of an instrument to measure moral distress within the Australian residential and community care environments. *International Journal of Older People Nursing* 12 (2): e12144. https://doi.org/10.1111/opn.12144.

Cameron, F., and S. Brownie. 2010. Enhancing resilience in registered aged care nurses. *Australasian Journal on Ageing* 29 (2): 66–71. https://doi.org/10.1111/j.1741-6612.2009.00416.x.

Canadian Nurses Association. 2008. *Code of ethics for registered nurses*. Ottawa, ON: Author.

Carnevale, F., B. Delogu, A. Bagnasco, and L. Sasso. 2018. Correctional nursing in Liguria, Italy: examining the ethical challenges. *Journal of Preventive Medicine and Hygiene* 59 (4): E315–E322. https://doi.org/10.15167/2421-4248/jpmh2018.59.4.928

Edwards, M.P., S.E. McClement, and L.R. Read. 2013. Nurses' responses to initial moral distress in long-term care. *Journal of Bioethical Inquiry* 10 (3): 325–336. https://doi.org/10.1007/s11673-013-9463-6.

Fortney, L., C. Luchterhand, L. Zakletskaia, A. Zgierska, and D. Rakel. 2013. Abbreviated mindfulness intervention for job satisfaction, quality of life, and compassion in primary care clinicians: A pilot study. *The Annals of Family Medicine* 11 (5): 412–420. https://doi.org/10.1370/afm.1511.

Fry, S.T., R.M. Harvey, A.C. Hurley, and B.J. Foley. 2002. Development of a model of moral distress in military nursing. *Nursing Ethics* 9 (4): 373–387. https://doi.org/10.1191/0969733002ne522oa.

Ganske, K.M. 2010. Moral distress in academia. *OJIN: The Online Journal of Issues in Nursing* 15 (3): 6. https://doi.org/10.3912/OJIN.Vol15No03Man06

Hamric, A.B. 2001. Reflections on being in the middle. *Nursing Outlook* 49 (6): 254–257. https://doi.org/10.1067/mno.2001.120247.

Hamric, A.B. 2012. Empirical research on moral distress: issues, challenges, and opportunities. *HEC Forum* 24 (1): 39–49). https://doi.org/10.1007/s10730-012-9177-x

Hardingham, L.B. 2004. Integrity and moral residue: nurses as participants in a moral community. *Nursing Philosophy* 5 (2): 127–134. https://doi.org/10.1111.j.1466-769X.2004.00160.x

Holloway, D. 2004. Ethical dilemmas in community learning disabilities nursing. *Journal of Learning Disabilities* 8 (3): 283–298. https://doi.org/10.1177/1469004704044968.

Jameton, A. 1984. *Nursing practice: The ethical issues*. Engelwood Cliffs, NJ: Prentice Hall.

Jameton, A. 1993. Dilemmas of moral distress: Moral responsibility and nursing practice. *AWHONN's Clinical Issues in Perinatal and Women's Health Nursing* 4: 542–551.

Jessiman, K. (2008). *Everyday ethics in case management: Experiences of moral distress by professionals in a community health care setting* (Unpublished master's thesis). Lakehead University, Thunder Bay, ON, Canada. Retrieved from https://knowledgecommons.lakeheadu.ca:7070/bitstream/handle/2453/3862/JessimanK2008m-1b.pdf?sequence=1

Kada, O., and T. Lesnik. 2019. Facets of moral distress in nusing homes: A qualitative study with examined registered nurses. *Zeitschrift fur Gerontologie und Geriatrie* 52: 743–750. https://doi.org/10/1007/s00391-019001645-w

Karlsson, I., S.L. Ekman, and I. Fagerberg. 2009. A difficult mission to work as a nurse in a residential care home–some registered nurses' experiences of their work situation. *Scandinavian Journal of Caring Sciences* 23 (2): 265–273. https://doi.org/10.1111/j.1471-6712.2008.00616.x.

Kenny, N.P., S.B. Sherwin, and F.E. Baylis. 2010. Re-visioning public health ethics: A relational perspective. *Canadian Journal of Public Health* 101 (1): 9–11. https://doi.org/10.1007/BF0340 5552.

Kuehner, C.A. 2013. My military: A navy nurse practitioner's perspective on military culture and joining forces for veteran health. *Journal of the American Academy of Nurse Practitioners* 25 (2): 77–83. https://doi.org/10.1111/j.1745-7599.2012.00810.x.

Laabs, C.A. 2005. Moral problems and distress among nurse practitioners in primary care. *Journal of the American Academy of Nurse Practitioners* 17: 76–84. https://doi.org/10.1111/j.1041-2972. 2005.00014.x.

Ladd, R.E., L. Pasquerella, and S. Smith. 2000. What to do when the end is near: Ethical issues in home health care nursing. *Public Health Nursing* 17 (2): 103–110. https://doi.org/10.1046/j. 1525-1446.2000.00103.x.

Lazzari, T., S. Terzoni, A. Destrebecq, L. Meani, L. Bonetti, and P. Ferrara. 2020. Moral distress in correctional nurses: A national survey. *Nursing Ethics* 27 (1): 40–52. https://doi.org/10.1177/ 0969733019834976.

Liaschenko, J. 2001. Nursing work, housekeeping issues, and the moral geography of home care. *Aging: Caring for our elders*. Dordrecht, Netherlands: Springer.

MacPhail, S. 2001. Ethical awareness for community-care nurses. *Touch: The Provincial Health Ethics Network Newsletter* 3 (11): 1–2.

Magnusson, A., and K. Lützén. 1999. Intrusion into patient privacy: A moral concern in the home care of persons with chronic mental illness. *Nursing Ethics* 6: 399–410.

Malone, R.E. 2003. Distal nursing. *Social Science & Medicine* 56 (11): 2317–2326. https://doi.org/ 10.1016/S0277-9536(02)00230-7

McMillan, N. (2020). Moral distress in residential child care. *Ethics and Social Welfare* 1–13. https://doi.org/10.1080/17496535.2019.1709878

Meaney, M. 2002. Moral distress: I just can't take it anymore! *The Case Manager* 13 (3): 32–33.

Minore, B., M. Boone, M. Katt, P. Kinch, S. Birch, and C. Mushquash. 2005. The effects of nursing turnover on continuity of care in isolated First Nation communities. *Canadian Journal of Nursing Research Archive* 37 (1): 86–100.

Oberle, K., and S. Tenove. 2000. Ethical issues in public health nursing. *Nursing Ethics* 7 (5): 425–438. https://doi.org/10.1177/096973300000700507.

Peter, E., and J. Liaschenko. 2004. Perils of proximity: a spatiotemporal analysis of moral distress and moral ambiguity. *Nursing Inquiry* 11 (4): 218–225. https://doi.org/10.1111/j.1440-1800. 2004.00236.x.

Pijl-Zieber, E.M., O. Awosoga, S. Spenceley, B. Hagen, B. Hall, and J. Lapins. 2018. Caring in the wake of the rising tide: Moral distress in residential nursing care of people living with dementia. *Dementia* 17 (3): 315–336. https://doi.org/10.1177/1471301216645214

Powers, R. 2003. Boundary crossings and boundary violations: Is there a difference? *Psychology of Women Quarterly* 27 (3): 277. https://doi.org/10.1111/1471-6402.00107_7.

Powers, M., and R. Faden. 2006. *Social justice: The moral foundations of public health and health policy*. New York, NY: Oxford University Press.

Public Health Agency of Canada. 2008. *Core competencies for public health in Canada:* Release 1.0. Ottawa, ON: Author. Retrieved from https://www.canada.ca/content/dam/phac-aspc/doc uments/services/public-health-practice/skills-online/core-competencies-public-health-canada/ cc-manual-eng090407.pdf

Racher, F.E. 2007. The evolution of ethics for community practice. *Journal of Community Health Nursing* 24 (1): 65–76. https://doi.org/10.1080/07370010709336586.

Ritchie, V., T. O'Rourke, and S. Stahlke. 2018. Nurse practitioners' experiences of moral distress in the continuing care setting. *The Journal for Nurse Practitioners* 14 (10): 745–752. https://doi. org/10.1016/j.nurpra.2018.07.021.

Rosenkoetter, M. 1983. A code of ethics for nurse educators. *Nursing Outlook* 31: 288.

Rosenkoetter, M.M., and J.A. Milstead. 2010. A code of ethics for nurse educators: Revised. *Nursing Ethics* 17 (1): 137–139. https://doi.org/10.1177/0969733009350946.

Rutenberg, C., and K. Oberle. 2008. Ethics in telehealth nursing practice. *Home Health Care Management & Practice* 20 (4): 342–348. https://doi.org/10.1177/1084822307310766.

Schoot, T., I. Proot, M. Legius, R. ter Meulen, and L. de Witte. 2006. Client-centred home care: Balancing between competing responsibilities. *Clinical Nursing Research* 15 (4): 231–254. https://doi.org/10.1177/1054773806291845.

Singer, M. 2007. *The untethered soul: The journey beyond yourself*. Oakland, California: New Harbinger Publications.

Singh, M.D., F.B. Pilkington, and L. Patrick. 2014. Empowerment and mentoring in nursing academia. *International Journal of Nursing Education Scholarship* 11 (1): 101–111. https://doi.org/10.1515/ijnes-2013-0070.

Spenceley, S., S. Caspar, and E.M. Pijl. 2019. Mitigating moral distress in dementia care: Implications for leaders in the residential care sector. *World Health Population* 18 (1): 47–60. https://doi.org/10.12927/whp2019.26059

Spenceley, S., C.S. Witcher, B. Hagen, B. Hall, and A. Kardolus-Wilson. 2017. Sources of moral distress for nursing staff providing care to residents with dementia. *Dementia* 16 (7): 815–834. https://doi.org/10.1177/1471301215618108.

Stulginski, M.M. 1993. Nurses' home health experience part I: The practice setting. *Nursing and Health Care* 14 (8): 402–407.

Tippit, M.P., N. Ard, J.R. Kline, J. Tilghman, B. Chamberlain, and G. Meagher. 2009. Creating environments that foster academic integrity. *Nursing Education Perspectives* 30 (4): 239–244.

Varcoe, C., G. Doane, B. Pauly, P. Rodney, J.L. Storch, K. Mahoney, G. McPherson, H. Brown, and R. Starzomski. 2004. Ethical practice in nursing: Working the in-betweens. *Journal of Advanced Nursing* 45 (3): 316–325.

Viehbeck, S.M., R. Melnychuk, C.W. McDougall, H. Greenwood, and N.C. Edwards. 2011. Population and public health ethics in Canada: A snapshot of current national initiatives and future issues. *Canadian Journal of Public Health* 102 (6): 410–413. https://doi.org/10.1007/BF0340 4188.

Vinje, H.F., and M.B. Mittelmark. 2008. Community nurses who thrive: The critical role of job engagement in the face of adversity. *Journal for Nurses in Professional Development* 24 (5): 195–202. https://doi.org/10.1097/01.NND.0000320695.16511.08

Weijer, C., A. Skelton, and S. Brennan. 2013. *Bioethics in Canada*. Don Mills, ON, Canada: Oxford University Press Canada.

Williamson, K. 2007. Home health care nurses' perceptions of empowerment. *Journal of Community Health Nursing* 24 (3): 133–153. https://doi.org/10.1080/07370010701429512.

Wojtowicz, B., and B. Hagen. 2014. A guest in the house: Nursing instructors' experiences of the moral distress felt by students during inpatient psychiatric clinical rotations. *International Journal of Nursing Education Scholarship* 11 (1): 121–128. https://doi.org/10.1515/ijnes-2013-0086.

Wojtowicz, B., B. Hagen, and C. Van Daalen-Smith. 2014. No place to turn: Nursing students' experiences of moral distress in mental health settings. *International Journal of Mental Health Nursing* 23 (3): 257–264. https://doi.org/10.1111/inm.12043.

Young, A., K. Froggatt, and S.G. Brearley. 2017. 'Powerlessness' or 'doing the right thing'–Moral distress among nursing home staff caring for residents at the end of life: An interpretive descriptive study. *Palliative Medicine* 31 (9): 853–860. https://doi.org/10.1177/0269216316682894.

Chapter 6
Personal and Professional Identities

Abstract A re-examination of traditional approaches to foundational discourses in health care ethics is duly needed. In contemporary western societies, there is ongoing attention to patient-centred and family-involved care. However, an additional commitment to the needs of carers is missing. Ethics is about who you are, what you do, and who you become as a result of the processes and outcomes of your choices and experiences. Accepting this statement will require a new ethic; one that acknowledges and supports the inherent risks and suffering that the valuable work of caring requires. A new ethic with a compassionate approach to all those in the circle of care and at many different levels. This chapter speaks to the health care professional as an individual, as a wounded healer (past, present and future), and it approaches complex issues from a relational perspective.

Keywords Caring · Moral sensitivity · Empathy · Compassion · Professional dissonance · Moral identity

6.1 Introduction

Caring too much is not the root of the problem for moral distress, just as caring less is not going to promote protection from or healing after moral distress experiences regardless of the practice setting. Writing this chapter leads me to recall a conversation with a nurse manager that occurred years ago when I was a newly graduated registered nurse at a busy, hospital in-patient unit. It is a story that has remained stuck in my mind.

It was time for my annual employee performance appraisal. The nurse manager asked me what my greatest weakness was that I felt the need to work on in the year to come. I answered without hesitation that I cared too much about the patients and my work on the unit. I really felt badly about this. Honesty, I took my work home (in my head) after every shift and thought about patients on my days off. I believed that maybe I needed to toughen up and that someone might have mentioned something to her. Better that I call it out first. She said, '*When you stop caring, you should quit nursing.*' She said this matter-of-factly and without room for questions. It seemed

© Springer Nature Switzerland AG 2020   87
K. Jones-Bonofiglio, *Health Care Ethics through the Lens of Moral Distress*,
The International Library of Bioethics 82,
https://doi.org/10.1007/978-3-030-56156-7_6

that she was telling me that a nurse who does not care is not really a nurse. This conversation, long-held in my mind, begs the question of whether nursing is about who I am, what I do, or a bit of both? And, what does it mean to 'care too much'? I will explore these questions further in this chapter and later in this book.

Let's begin with a discussion about professional identity for nurses and connections to ethics in nursing practice. Since World War I, nursing has been a regulated profession in Canada (Keatings and Smith 2010). Over time nursing codes of ethics have evolved, as nursing as a profession has evolved from having a limited role, such as being the physicians' handmaiden, to a more fulsome scope of professional nursing practice (Viens 1989). The first code of ethics for nurses was established by the International Council of Nurses (ICN) in the early 1950s (ICN 1953). The code was only a single page long and contained just 14 sentences. It mapped out four fundamental nursing responsibilities: (1) promoting health; (2) preventing illness; (3) restoring health; and, (4) alleviating suffering. This code focused largely on the nurse, her responsibilities, and maintaining professional conduct. The word '*patient*' was used only once.

The Canadian Nurses Association (CNA) has been a longstanding member of the ICN. As such, in 1954 the CNA adopted the ICN's code of ethics for use by Canadian nurses. It was not until 1980 that the CNA created its own code of ethics specific to Canadian nurses, as a statement of values and commitments. Seeking to be far from the focus on virtue and etiquette, modern nursing codes were designed as frameworks for integrating nursing ethics theory into practice, guiding moral reflection, and supporting ethical decision making (Keatings and Smith 2010). Such codes provide a systematic and consistent approach to analyzing and articulating ethical aspects of nursing practice (Davis et al. 2010). The current version of the CNA's code of ethics for registered nurses (CNA 2017) is 60 pages long and it is made up of two parts that include: (1) values and ethical responsibilities; and, (2) ethical endeavours (e.g., social justice approaches, accountability to relationships, advocacy). Almost all points relate specifically to '*persons receiving care*'. Here I will argue that something has been lost. Although, having an ethics guidebook and following these 'rules of thumb' should be a source of pride and satisfaction for nurses, it has been suggested that professional codes of ethics are deficient and may actually cause more confusion, apathy, and unethical behaviours (Pattison 2001). How does this affect nurses?

A Brazilian study of critical care nurses explored the concept of nursing virtues and how constraints related to expressing one's self as a 'good nurse' may lead to moral distress experiences and a sense of 'invisibility of the self' (Caram et al. 2019). This ideal of the good nurse has a historical shadow with both military and religious aspects; from selfless stoic to angel of mercy. There are other images of nursing that are not so heroic. Nurses have been depicted as sex objects in media and film, such as the head nurse character of Major Margaret 'Hot Lips' Houlihan in the 1970s classic television show M.A.S.H. Nurses have also been depicted as heartless tyrants, such as Nurse Ratched in the book and film One Flew Over the Cuckoo's Nest. I propose that other health care professions do not have such stigmatizing stereotypes as deeply engrained into society's views about their professional identity.

Canadian sociology professor (emeritus) Arthur Frank (2004) asserts that ethics "needs to shift its orientation from decisions to identities-that is, who I become as a result of making this decision" (p. 357). This is an important distinction, as ethics in nursing is not only about the ethical choices that are made. There is a much bigger picture to consider. Nursing ethics extends to the relational outcomes of ethical decisions and to the impacts that ripple out from individuals to interpersonal, institutional or agency, community, and/or societal levels. The potential ripple effect of nurses' ethical decision making can be far reaching and complex.

A relational context for ethics and ethical decision making is found in relational ethics theory. This theory describes "that our ethical understandings are formed in, and emerge from, our relationships with others" (Oberle and Bouchal 2009, p. 40). As well, being in relationship with another, results in a sharing of that person's experiences (Maeve 1998). Relational ethics requires consideration of multiple contexts within and between individuals, societies, as well as power inequities (Rodney et al. 2013). It is within the relational space where morality is enacted and the effects can potentially be felt at all levels of health care practice (Bergum 2013).

As described by Canadian nurse scholars Diane Kunyk and Wendy Austin (2011), "nursing provides rewarding and enriching careers for its members but, at the same time, often proves to be demanding, stressful and isolating" (p. 381). Thus, it is important to situate nursing into the ethical dialogue of health care practice. This needs to be done by nurses for nurses. To take our understanding of ethics to a deeper level, ethics can be considered as "a way of being, a process of enactment" (Varcoe et al. 2004, p. 319). From a relational perspective, this process of enactment requires individual nurses to critically reflect on the choices available to self and others, justifications for those options, and how to best respond (Hardingham 2004). Therefore, nursing ethics can be described as what you do, how you do it, who you are, and who you become, during and after experiences, all of which encompasses a more complex process than simply applying moral philosophy.

6.2 Being Human

German scholar Daniel Tigard (2019) suggests that moral distress can be a powerful source of knowledge and understanding about our true nature as caring and engaged human beings. Our emotional responses, to ethical issues that transgress the things that we care about, can be a source of affirmation of our moral identity as a person and as a professional. Instead of seeking to alleviate negative responses, individuals may consider upholding these responses as the mark of an honorable moral character. Further, circumstances that put our personal and professional values to the test, can serve as important learning experiences and provide opportunities for growth.

In 2019, I attended a conference hosted by the International Academy of Law and Mental Health (IALMH). There I was introduced to the reality that robot judges are currently being used in China. The learned presenter asked the audience, "*does a judge need to be human?*" I immediately wondered about this for nursing. My gut reaction was to respond that, of course, a non-human nurse would be impossible, but I actually did not readily have an answer as to why. Further research into this presented me with the fact that over 35 million service robots (e.g., carebots) are in use worldwide (van Wynsberghe 2016). Also, researchers at Carnegie Mellon University (n.d.) in Pittsburg have created 'Flo the Nursebot' for assisting elderly clients with reminders, manipulating objects, videoconferencing (telepresence), data collection (e.g., vital signs), and social interaction. Yes, that is correct, also for social interaction (i.e., relationships). Researchers at Carnegie Mellon are leading studies of human-robotic interactions by meshing cognitive psychology with machine learning to help robots anticipate and deliver on human needs and desires (Henninger 2020).

I would like to comment that, in the future, health care providers are going to have to start thinking about what it really means to be human and how delivering the self (Harris and Griffin 2015) in health care practice really matters. But the time is actually now. Right now. We are living in the midst of this reality. In fact, health care providers need to be able to articulate, quantify, and justify the actual value of caring, compassion, and human connection. Does this sound crazy?

In 2019, two American physicians did just that and wrote a book, *Compassionomics: The revolutionary scientific evidence that caring makes a difference* (Trzeciak et al. 2019). The book is written in the context of achieving better outcomes, driving revenues, cutting costs, and as the antidote to burnout. Essentially the authors describe that caring and compassion (the human element) has an excellent return-on-investment (ROI), largely because it appears to require little-to-nothing from systems and organizations and everything from individual care providers as moral agents. Perhaps there is a role for humans after all.

6.3 Conflicting Obligations

Humans have the potential to take on unique roles as moral agents. Before enacting moral agency, health care providers need to be consciously aware that an ethical issue is in their midst. Moral sensitivity (also known as ethical attunement) allows care providers to become aware of and recognize the variable dimensions of an ethical issue (Lützen et al. 2010; Lützén et al. 2000; Milliken 2018; Rushton et al. 2013). For nurses, ethics requires an understanding of both the virtues of a nurse and the obligations of nursing (Oberle and Bouchal 2009).

From the mid 1800s onward, nursing was considered to be a vocation (White 2002) and early nurse leaders, such as Florence Nightingale, promoted the need for moral education based on Aristotle's views of virtue ethics and Christian moral principles (Sellman 1997). Thus, modern nursing ethics, as defined by Keatings and Smith (2010), involves the moral questions within nursing practice and establishes nurses as moral agents whose character and relationships are the foundation. This burden of moral work comes from the professional judgments that nurses make that affect peoples' lives, involving relationships and conflicts of values, duties, and priorities (Davis et al. 2010).

Competing interests in nursing, also known as conflicting obligations (Provis and Stack 2004), are born of fiduciary (legal; Austin 2012) and covenantal (sacred trust) relationships with multiple stakeholders (Meaney 2002). Health care providers often feel called to act with individual and collective moral obligations at stake. Within the context of this web of priorities, dual loyalties may exist (Appel 2014). The existence of these conflicts recognizes the complex relationality of health care professions and their practice realities. The work of American philosopher John Caputo urges us to seek further understanding by asking more questions rather than looking for prescribed answers and to accept that our obligations are both externally and internally motivated (Caputo 1987, 1993).

6.4 Dissonance

Much of the burden of the moral work of nurses is actually hidden and therefore achieving a moral equilibrium is difficult, if not impossible. Nursing scholars have written about the 'invisibility' of the work of nurses and their valuable situated moral knowledge (Bjorklund 2004; [artificial personhood] Liaschenko 1995; Rodney and Varcoe 2001). Thus nurses and other caring professionals must devise strategies to cope. Cognitive dissonance is one mechanism used to manage the pressure of values conflicts. Often explained using Aesop's famous fable of the fox and the grapes, this strategy involves emotional self-soothing through narrative reframing of one's thoughts to reduce frustration. And so, let me tell you the story in the way that I have heard it told:

> As the story goes, one day a bright and energetic fox came across a fully ripe bunch of juicy grapes that were growing on a vine just out of his reach. Over the next few days and nights he continued to long for the grapes and came to deeply desire to be able to taste them. Oh, how splendid that would be! As he pictured in his mind how wonderful they would taste, his eyes would grow wide and saliva would fill up his mouth. However, his many efforts to reach the grapes seemed to always be in vain. Eventually the fox grew very tired of his lack of success. Finally, the fox came to the realization that he would likely never obtain the beautiful and delicious grapes. He became very unhappy. He paced and shouted. He cried many tears. Eventually, he stood up and wiped his face. He then spun a detailed narrative, to himself, that the grapes were likely to be very sour. Yes, they were most definitely going to be distasteful. Probably, they would be wormy in the middle. Therefore, he had no reason to continue to want them. In fact, he wondered why he had ever thought about tasting them in the first place. And so, he walked away from the bunch of grapes smiling and feeling glad that he had been so clever to have saved himself from such a disappointment. This bright and energetic fox, once filled with motivation and desire, had quelled his unmet expectations with an objective rationalization that gave him comfort and allowed the world to make sense once again.

We can see this coping mechanism in action among health care providers today, although usually in much more subtle ways. Attention has shifted to prioritizing the completion of doable tasks with a focus on moving patients through the system by, first and foremost, addressing a primary diagnosis; not a unique individual with a

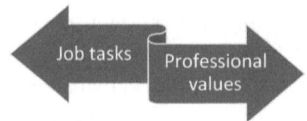

Fig. 6.1 Professional dissonance

complex life and set of circumstances. In essence, this is setting one's course on the potentially achievable goals and setting everything else to the side as much as possible. Canadian nursing scholars Gweneth Hartrick Doane and Colleen Varcoe (2007) agree that this occurs, with nurses seeking some semblance of control by pulling close concrete tasks and pulling away from the unfixable. It makes sense, and one could even argue that this is a necessary response of self-protection, due to the changes that contemporary health care professionals have endured (and continue to endure). Some nurses even wax nostalgic for the 'good old days' when there was time for patient back rubs and perfect hospital corners on bedsheets. Today, the focus is largely on survival- not just for patients, but for health care providers too.

Professional dissonance is a related term, where there is conflict between a job task and one's professional values. In an American study of over 300 social workers, connections between job tasks and personal characteristics related to professional dissonance were explored (Taylor and Bentley 2005). Findings indicated that individuals' characteristics were most closely related to levels of dissonance. Study participants with the highest levels of dissonance were male social workers with many years of practice (Fig. 6.1).

Opportunities to take pride in one's work and to have a sense of accomplishment are key aspects of experiencing work satisfaction and professional fulfillment. The coping mechanism of reducing one's complex professional role and responsibilities to a list of doable tasks, can lead health care providers to feel a semblance of control in a chaotic environment. Other means of survival include devising strategies of cutting corners, work-arounds (Debono et al. 2013), bending the rules (Collins 2012; Hutchinson and Collins 2004), rationing care (Jones et al. 2015; Papastavrou et al. 2014), and emotional distancing. Meanwhile, at the end of each shift when quantity may have prevailed, quality got lost in the mix.

6.5 Moral Identity

Moral identity can be defined as the basing of one's identity and self-concept on consistently maintained positive moral values (Johnston, Sherman, and Grusec 2013). Moral integrity can be further defined as adherence to these values (Corley 2002). Therefore, if one is able to consistently adhere to these values, the integrity of one's moral identity may be successfully maintained. If there is a perceived compromise of moral values, then erosion may occur (Epstein and Hamric 2009). The situation of being unable to promote and maintain positive moral values may lead to less than optimal outcomes for patients and for health care providers.

When values are not upheld and patients suffer, nurses are there to witness it. Therefore, it is the proximity to suffering that allows nurses to become attuned to the ethical issues in their practice. These affective dimensions of nursing practice, often referred to as emotional labour (Funk, Peters, and Roger 2017), can make ethical decision making much more complicated (Jameton 2013), however it also makes it possible in the first place. Reaching back to the original code of nursing ethics (ICN 1953), nurses were tasked with the following core ethical responsibilities to life, health, and the reduction of suffering, as noted in Table 6.1.

Each of the 14 points in this code are still highly relevant to nursing almost seven decades later. Attention to these historical ethics guide posts, which focused directly on the nurse and indirectly on the patient, highlights areas where nurses can

Table 6.1 Nurses' ethical responsibilities from ICN (1953)

1	The fundamental responsibility of the nurse is threefold: to conserve life, to alleviate suffering, and to promote health.
2	The nurse must maintain at all times the highest standards of nursing care and of professional conduct.
3	The nurse must not only be well prepared to practice but must maintain her knowledge and skill at a consistently high level.
4	The religious beliefs of a patient must be respected.
5	Nurses hold in confidence all personal information entrusted to them.
6	A nurse recognizes not only the responsibilities but the limitations of her or his professional functions; recommends or gives medical treatment without medical orders only in emergencies and reports such action to a physician at the earliest possible moment.
7	The nurse is under an obligation to carry out the physician's orders intelligently and loyally and to refuse to participate in unethical procedures.
8	The nurse sustains confidence in the physician and other members of the health team; incompetence or unethical conduct of associates should be exposed, but only to the proper authority.
9	A nurse is entitled to just remuneration and accepts only such compensation as the contract, actual or implied, provides.
10	Nurses do not permit their names to be used in connection with the advertisement of products or with any other forms of self-advertisement.
11	The nurse cooperates with and maintains harmonious relationships with members of other professions and with her or his nursing colleagues.
12	The nurse in private life adheres to standards of personal ethics which reflect credit upon the profession.
13	In personal conduct nurses should not knowingly disregard the accepted patterns of behavior of the community in which they live and work.
14	A nurse should participate and share responsibility with other citizens and other health professions in promoting efforts to meet the health needs of the public- local, state, national, and international.

strengthen and maintain a strong moral identity. After all, a person with a strong moral identity is prepared to offer two resources: knowledge and action. Knowledge can be sourced from education and/or experience and sometimes it is a complex mixture of the two. Action involves the courage to act, competency (the skill to do it), and commitment to see it through (Maeve 1998). Now we come to the crux of the problem. It is precisely when moral agency is constrained (action is thwarted) that moral distress experiences begin to erode nurses' professional and/or personal moral identities.

Many authors have argued that it is the moral component of nursing practice that predisposes nurses to experience moral distress related to the ethical challenges that impact one's values and beliefs (Hamric 2012; Lützén and Kvist 2012), sense of power (Epstein and Hamric 2009), and integrity (Corley 2002). However, if nursing practice has an inherently moral component, then would it not make sense that moral distress must be an unavoidable part of nursing practice?

I argue that some experiences of moral distress are unavoidable because moral distress is about navigating uncertainty and asking (sometimes difficult) questions. Could moral distress ever be completely avoided? Yes, but that would need to occur in a health care environment where the ethical aspects of practice were deemed to be of no immediate importance. No moral distress, ever, would mean that ethical questions are not being raised, addressed, or resolved. It would mean that individuals would choose to silence the call of their ethical beliefs and ignore their moral compass.

Recap of Concept: Nurses' Moral Identity
- ✓ Positive moral values
- ✓ High standards
- ✓ Knowledge
- ✓ Competency and skills
- ✓ Shared responsibilities
- ✓ Respect and trust
- ✓ Confidence and cooperation
- ✓ Personal ethical comportment
- ✓ Commitment and courage to act.

6.6 Cost of Caring

Truly caring requires vulnerability, emotional engagement, giving of self, and a willingness to be in the present moment with another. Ethical concepts such as caring and caring in therapeutic relationships must be considered in this discussion. Since the 1980s an ethics of caring has been discussed in the nursing literature (Davis et al. 2010), although it was certainly present before that even if not fully and academically articulated. There is an argument for caring as a primary ethical value, especially in professional nursing practice, in order to create a foundation of a wider moral framework for ethics to be established (Held 2006; Herring 2013).

In declaring "…caring relation as ethically basic…" (Noddings 2002, p. 3), American feminist philosopher Nel Noddings urges carers to focus on "…how we meet the other morally…" (p. 5). In this statement, she draws attention to the fact that caring practices are about more than just making a rational, objective ethical decision and re-applying that decision to similar ethical issues in the future. Ethics is about the unique contexts of each individual caring relationship, how decisions are carried out, and the relational outcomes that flow from those actions, behaviours, and attitudes. This can be an uncomfortable place for moral philosophy (and for health care providers) to settle as it may be considered to be far too subjective or 'fluffy'.

Notwithstanding, it is the caring practices and relational processes that create a moral dimension to the work and professional identity of nurses. "Thus a morally good nurse is actively concerned with fostering well-being through a caring relationship in the context of attentive, efficient, and effective nursing practice" (Oberle and Bouchal 2009, p. 44). The 'good nurse' is able to achieve the patient's best interests and optimal professional outcomes. Therefore, it is important that professional caring, with both positive and negative (e.g., moral distress) experiences, be understood in a relational context shaped by the interconnections between the individuals involved, influenced by professional obligations, and impacted by the multiple dynamics that exist within.

6.7 Cost of Not Caring

American nursing theorist, Jean Watson (1988), describes nursing as an intersubjective experience which occurs due to individuals' cognitive and emotional perspectives. While there is a science of caring in nursing, there is also an art of nursing that challenges attempts to define, measure, and evaluate the costs and the benefits of caring. I once had a professor in graduate school who stated, 'if it cannot be measured, it is not worth researching'. I was stopped in my tracks and thought, "Are there not important factors in this world that researchers may not be able to quantify?" Perhaps, this is the space that is called to be occupied by those with skills in ethics. Those who are more comfortable wading into the unknown, while still acknowledging known facts and able to contextualize statistics.

What does it cost health care providers to not care? Over time, humans have developed mirror neurons in their brains that light up pain centres when empathy for another is experienced. The brain does not differentiate the pain of another from pain of the self. Therefore the pathway to apathy goes against our hardwired human biology, unless of course some type of psychopathology exists. The cost of apathy is two fold. It costs the individual and likely ripples out to affect those around them. Further being physically present and bearing witness to suffering and not emotionally engaging requires dissociation or disembodiment from the present moment (Maeve 1998). This is the realm of poor mental health and mental illness.

6.8 Suffering Without Meaning

The concept of a 'wounded healer' has been proposed by great minds such as Carl Jung (psychologist), Henri Nouwen (theologian), and Joan Halifax (anthropologist/Zen Buddist teacher [Roshi]). At times, the nursing ethics literature speaks to the moral suffering of nurses and portrays nurses as victims (McCarthy and Deady 2008). Nurses themselves usually vehemently reject being cast as victims or even the thought of sounding like a victim (Magnussen 2019). Perhaps this assumption about nurses as victims comes from a traditional 'culture of silence' (Ross et al. 2018; Verhezen 2010). I propose that this cultural of silence has left nurses feeling isolated and unsupported in their practice by others (e.g., team members, organizations, systems) and by the profession of nursing itself. In essence, silence has caused nurses to suffer without others truly understanding the meaning of this suffering. In a culture of silence, there is a sense of helplessness (in trying to explain it) and a sense of hopelessness (that it could ever be fully understood or changed). Perhaps we can see this fall out in terms of nurses' responses to moral distress and tendencies to simply withdraw (Epstein and Hamric 2009; Hamric 2012).

The potential for nurses to be perceived (or to perceive themselves) as victims calls to mind work done by Karpman (1968) on the 'drama triangle' and the valuable opportunity for nurses to disentangle themselves from dependence and move toward empowerment (see Fig. 6.2 to understand the dynamics of the triad of the drama triangle). In order to do this, individuals must recognize their own 'script', move toward self-reliance by adopting a responsible attitude, make behaviour changes where necessary, and become the author of one's fate (Shmelev 2015). This is the ability to tell one's story differently.

One strategy to do this is through 'workplace voice'. In an integrative review of the literature on workplace voice, the authors found a variety of options for being heard, including suggestion systems, prosocial voice, informal complaints, grievances, and whistleblowing (Klaas et al. 2012). Considerations for navigating safety and risk during the use of workplace voice are noted as formality, focus, and identifiability.

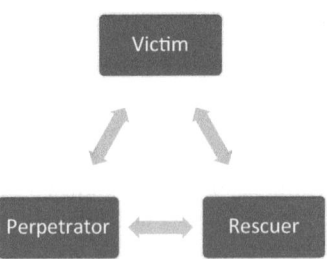

Fig. 6.2 Dynamic components of the drama triangle, based on the work of Karpman (1968)

6.9 Conclusion

Some nursing scholars see the current state of flux of contemporary health care environments as an emerging opportunity for nurses to work together, advocate, and create positive changes (Tomajan 2012), for individual nurses and for the profession itself. A caring identity, as a person and as a professional, is often central to individual health care provider's sense of moral integrity and moral agency. Historical discourses on health care ethics need to be reviewed. It is important to recognize that the current codes of ethics have evolved from what went before them. Older codes and literature on professional ethics should not be discounted entirely, but rather explored to understand the cores values and moral foundations that speak to and have pragmatically supported ethical health care practice across time. Further, extensive review of current codes of ethics need to occur. Are they functional or are they actually contributing to dysfunction? There is a cost to an ethical standard that cannot be consistently upheld and that cost is human suffering. Individual strategies to address moral distress are important to give health care providers options to build capacities for knowledge, attitudes, beliefs, and skills that will help them to navigate ethics in their everyday practice. While change does need to occur on many levels, the voices that call for change need to call out, loud and clear. And the questions, that are sometimes whispered, need to be heard and responded to:

Is nursing about who I am, what I do, or a bit of both?
What does it mean to 'care too much'?

References

Appel, J.M. 2014. Nonconsensual blood draws and dual loyalty: When bodily integrity conflicts with the public health. *Journal of Health Care Law & Policy* 17: 129–154.

Austin, W. 2012. Moral distress and the contemporary plight of health professionals. *HEC Forum* 24 (1): 27–38. https://doi.org/10.1007/s10730-012-9179-8.

Bergum, V. 2013. Relational ethics for health care. In *Toward a moral horizon: Nursing ethics for leadership and practice*, 2nd ed., ed. J.L. Storch, P. Rodney, and R. Starzomski, 127–142. Don Mills, ON: Pearson Canada Inc.

Bjorklund, P. 2004. Invisibility, moral knowledge and nursing work in the writings of Joan Liaschenko and Patricia Rodney. *Nursing Ethics* 11 (2): 110–121. https://doi.org/10.1191/096 9733004ne677oa.

Canadian Nurses Association (CNA). 2017. *Code of ethics for registered nurses*. Retrieved from https://cna-aiic.ca/~/media/cna/page-content/pdf-en/code-of-ethics-2017-edition-secure-intera ctive.pdf?la=en

Caputo, J. 1987. *Radical hermeneutics: Repetition, deconstruction and the hermeneutic project*. Bloomington, Indiana: Indiana University Press.

Caputo, J. 1993. *Against ethics: Contributions to a poetics of obligation with constant reference to deconstruction*. Bloomington, Indiana: Indiana University Press.

Caram, C.S., E. Peter, and M.J.M. Brito. 2019. Invisibility of the self: Reaching for the telos of nursing within a context of moral distress. *Nursing Inquiry* 26 (1): e12269. https://doi.org/10. 1111/nin.12269.

Carnegie Mellon University. n.d. Retrieved from https://www.cs.cmu.edu/~flo/press/cmu_99_12/nursebot.html

Collins, S.E. 2012. Rule bending by nurses: Environmental and personal drivers. *Journal of Nursing Law* 15 (1): 14–26. https://doi.org/10.1891/1073-7472.15.1.14.

Corley, M.C. 2002. Nurse moral distress: A proposed theory and research agenda. *Nursing Ethics* 9 (6): 636–650. https://doi.org/10.1191/0969733002ne557oa

Davis, A.J., M.D. Fowler, and M.A. Aroskar. 2010. *Ethical dilemmas and nursing practice*, 5th ed. Upper Saddle River, NJ: Pearson Education Inc.

Debono, D.S., D. Greenfield, J.F. Travaglia, J.C. Long, D. Black, J. Johnson, and J. Braithwaite. 2013. Nurses' workarounds in acute healthcare settings: a scoping review. *BMC Health Services Research* 13 (1): 175. https://doi.org/10.1186/1472-6963-13-175.

Epstein, E.G., and A.B. Hamric. 2009. Moral distress, moral residue, and the crescendo effect. *The Journal of Clinical Ethics* 20 (4): 330–342.

Frank, A.W. 2004. Ethics in medicine: Ethics as process and practice. *Internal Medicine* 34: 355–357. https://doi.org/10.1111/j.1445-5994.2004.00622.x.

Funk, L.M., S. Peters, and K.S. Roger. 2017. The emotional labor of personal grief in palliative care: Balancing caring and professional identities. *Qualitative Health Research* 27 (14): 2211–2221.

Hamric, A.B. 2012. Empirical research on moral distress: Issues, challenges, and opportunities. *Healthcare Ethics Forum* 24 (1): 39–49. https://doi.org/10.1007/s10730-012-9177-x.

Hardingham, L.B. 2004. Integrity and moral residue: Nurses as participants in a moral community. *Nursing Philosophy* 5 (2): 127–134. https://doi.org/10.1111/j.1466-769X.2004.00160.x.

Harris, C., and M.T.Q. Griffin. 2015. Nursing on empty: Compassion fatigue signs, symptoms, and system interventions. *Journal of Christian Nursing* 32 (2): 80–87. https://doi.org/10.1097/CNJ.0000000000000155.

Hartrick Doane, G., and C. Varcoe. 2007. Relational practice and nursing obligations. *Advances in Nursing Science* 30 (3): 192–205.

Held, V. 2006. *The ethics of care*. Oxford, UK: Oxford University Press.

Henninger, M. 2020. Shaping the future of human-robot interactions at Davos. *Carnegie Mellon University News*. Retrieved from https://www.cmu.edu/news/stories/archives/2020/january/world-economic-forum-admoni.html

Herring, J. 2013. Forging a relational approach: Best interests or human rights? *Medical Law International* 13 (1): 32–54. https://doi.org/10.1177/0968533213486542.

Hutchinson, S., and S.E. Collins. 2004. Leader interview: Nurses and bending the rules. *Creative Nursing* 9 (4): 4–8. https://doi.org/10.1891/1078-4535.9.4.4.

International Council of Nurses (ICN). 1953. *The international code of nursing ethics*. Retrieved from https://ethics.iit.edu/codes/ANA%201953.pdf

Jameton, A. 2013. A reflection on moral distress in nursing together with a current application of the concept. *Journal of Bioethical Inquiry* 10 (3): 297–308. https://doi.org/10.1007/s11673-013-9466-3.

Johnston, M.E., A. Sherman, and J.E. Grusec. 2013. Predicting moral outrage and religiosity with an implicit measure of moral identity. *Journal of Research in Personality* 47: 209–217. https://doi.org/10.1016/j.jrp.2013.01.006

Jones, T.L., P. Hamilton, and N. Murry. 2015. Unfinished nursing care, missed care, and implicitly rationed care: State of the science review. *International Journal of Nursing Studies* 52 (6): 1121–1137. https://doi.org/10.1016/j.ijnurstu.2015.02.012.

Karpman, S. 1968. Fairy tales and script drama analysis. *Transactional Analysis Bulletin* 7 (26): 39–43.

Keatings, M., and O. Smith. 2010. *Ethical & legal issues in Canadian nursing*, 3rd ed. Toronto, ON: Elsevier Canada.

Klaas, B.S., J.B. Olson-Buchanan, and A.K. Ward. The determinants of alternative forms of workplace voice: An integrative perspective. *Journal of Management* 38 (1): 314–345. https://doi.org/10.1177/0149206311423823

Kunyk, D., and W. Austin. 2011. Nursing under the influence: A relational ethics perspective. *Nursing Ethics* 19 (3): 380–389. https://doi.org/10.1177/0969733011406767.

Liaschenko, J. 1995. Artificial personhood: Nursing ethics in a medical world. *Nursing Ethics* 2 (3): 185–196. https://doi.org/10.1177/096973309500200302.

Lützen, K., T. Blom, B. Ewalds-Kvist, and S. Winch. 2010. Moral stress, moral climate and moral sensitivity among psychiatric professionals. *Nursing Ethics* 17 (2): 213–224. https://doi.org/10.1177/0969733009351951.

Lützén, K., A. Johansson, and G. Nordström. 2000. Moral sensitivity: Some differences between nurses and physicians. *Nursing Ethics* 7 (6): 520–530. https://doi.org/10.1177/096973300000700607.

Lützén, K., and B.E. Kvist. 2012. Moral distress: a comparative analysis of theoretical understandings and inter-related concepts. *HEC Forum* 24 (1): 13–25). https://doi.org/10.1007/s10730-012-9178-9

Maeve, K. 1998. Weaving a fabric of moral meaning: How nurses live with suffering and death. *Journal of Advanced Nursing* 27 (6): 1136–1142. https://doi.org/10.1046/j.1365-2648.1998.00622.x.

Magnussen, H. 2019. A response to Nico Nortjé's review of The Moral Work of Nursing. *Canadian Journal of Bioethics* 2 (2): 51. https://doi.org/10.7202/1058142ar.

McCarthy, J., and R. Deady. 2008. Moral distress reconsidered. *Nursing Ethics* 15 (2): 254–262. https://doi.org/10.1177/0969733007086023.

Meaney, M. 2002. Moral distress: I just can't take it anymore! *The Case Manager* 13 (3): 32–33. https://doi.org/10.1067/mcm.2002.124511.

Milliken, A. 2018. Nurse ethical sensitivity: An integrative review. *Nursing Ethics* 25 (3): 278–303. https://doi.org/10.1177/0969733016646155.

Noddings, N. 2002. *Educating moral people: A caring alternative to character education*. Chicago, Illinois: Teacher's College Press.

Oberle, K., and S.R. Bouchal. 2009. *Ethics in Canadian nursing practice: Navigating the journey*. Toronto, ON, Canada: Pearson Canada Inc.

Papastavrou, E., P. Andreou, and G. Efstathiou. 2014. Rationing of nursing care and nurse–patient outcomes: A systematic review of quantitative studies. *The International Journal of Health Planning and Management* 29 (1): 3–25. https://doi.org/10.1002/hpm.2160.

Pattison, S. 2001. Are nursing codes of practice ethical? *Nursing Ethics* 8 (1): 5–18. https://doi.org/10.1177/096973300100800103.

Provis, C., and S. Stack. 2004. Caring work, personal obligation and collective responsibility. *Nursing Ethics* 11 (1): 5–14. https://doi.org/10.1191/0969733004ne662oa.

Rodney, P., M. Burgess, B.M. Pauly, and J.C. Phillips. 2013. Our theoretical landscape: Complementary approaches to health care ethics. In *Toward a moral horizon: Nursing ethics for leadership and practice* (2nd ed.), eds. J.L. Storch, P. Rodney, and R. Starzomski, 84–106. Don Mills, ON: Pearson Canada Inc.

Rodney, P., and C. Varcoe. 2001. Towards ethical inquiry in the economic evaluation of nursing practice. *Canadian Journal of Nursing Research Archive* 33 (1): 35–57.

Ross, C.A., S.L. Jakubec, N.S. Berry, and V. Smye. 2018. "A two glass of wine shift": Dominant discourses and the social organization of nurses' substance use. *Global Qualitative Nursing Research* 5: 1–12. https://doi.org/10.1177/2333393618810655.

Rushton, C.H., A.W. Kaszniak, and J.S. Halifax. 2013. A framework for understanding moral distress among palliative care clinicians. *Journal of Palliative Medicine* 16 (9): 1074–1079. https://doi.org/10.1089/jpm.2012.0490.

Sellman, D. 1997. The virtues in the moral education of nurses: Florence Nightingale revisited. *Nursing Ethics* 4 (1): 3–11. https://doi.org/10.1177/096973309700400102.

Shmelev, I.M. 2015. Beyond the drama triangle: The overcoming self. *Psychology. Journal of Higher School of Economics* 12(2):133–149.

Taylor, M.F., and K.J. Bentley. 2005. Professional dissonance: Colliding values and job tasks in mental health practice. *Community Mental Health Journal* 41 (4): 469–480. https://doi.org/10.1007/s10597-005-5084-9.

Tigard, D.W. 2019. The positive value of moral distress. *Bioethics* 33 (5): 601–608. https://doi.org/10.1111/bioe.12564.

Tomajan, K. 2012. Advocating for nurses and nursing. *OJIN: The Online Journal of Issues in Nursing* 17(1): manuscript 4. https://doi.org/10.3912/OJIN.Vol17No01Man04

Trzeciak, S., A. Mazzarelli, and C. Booker. 2019. *Compassionomics: The revolutionary scientific evidence that caring makes a difference.* Pensacola, Florida: Studer Group.

van Wynsberghe, A. 2016. Service robots, care ethics, and design. *Ethics and Information Technology* 18 (4): 311–321. https://doi.org/10.1007/s10676-016-9409-x.

Varcoe, C., G. Doane, B. Pauly, P. Rodney, J.L. Storch, K. Mahoney, G. McPherson, H. Brown, and R. Starzomski. 2004. Ethical practice in nursing: Working the in-betweens. *Journal of Advanced Nursing* 45 (3): 316–325. https://doi.org/10.1046/j.1365-2648.2003.02892.x.

Verhezen, P. 2010. Giving voice in a culture of silence. From a culture of compliance to a culture of integrity. *Journal of Business Ethics* 96 (2): 187–206. https://doi.org/10.1007/s10551-010-0458-5

Viens, D.C. 1989. A history of nursing's code of ethics. *Nursing Outlook* 37 (1): 45–49.

Watson, J. 1988. *Nursing: Human science and human care.* New York, NY: National League for Nursing.

White, K. 2002. Nursing as vocation. *Nursing Ethics* 9 (3): 279–290. https://doi.org/10.1191/0969733002ne510oa.

Chapter 7
More Than Moral Distress

Abstract Nursing is an inherently moral endeavour. As such, the ability to be morally attuned to others situates nurses to be at increased risk for moral injury. This ties closely with experiences of moral distress, compassion fatigue, and burnout, but is something more. Nurses may enter the profession already wounded and seek to heal themselves through service to others. Nurses can be wounded by the work of nursing itself. To use the metaphor of a hero, there is little emphasis placed on attending to the needs of the hero. Psychological wounds, in particular, may go unacknowledged and unattended. Over time, a carer may begin to feel that they cannot make a difference or that they are actually contributing in some way, instead, to negative outcomes. This line of thinking can evolve into a deep-seated moral crisis.

Keywords Moral space · Moral ambiguity · Moral injury · Post traumatic growth

7.1 Introduction

The moral matrix (Lützén et al. 2003) of health care includes a paradox of overflow (e.g., patients, paperwork, technology, decisions, tasks, etc.) and underflow (e.g., resources, time, funding, front line staff, etc.). Today, health care is an art, a science, a service, and a business. Patients believe that they are more savvy than ever (thanks to doctor google), often with a list of expectations, and can demand time and patience that may not be readily available (Sethi and Salinas 2015).

Morality itself is also a paradox; self versus other, with various competing interests and obligations. However, moral action (see Fig. 7.1) is basically driven by three emotional tensions: (1) the need to feel/be moral; (2) the need to feel/appear moral to others; and, (3) the need to feel/believe the world is just (Haan 1989).

During moral distress experiences, the perception is that moral action has been blocked. Therefore one may expect potential impacts to one or more of these three areas: self, others, and/or world. Chapter six spoke to the impact on self in terms of both personal and professional identities. *Am I a good person? Am I a good nurse?* This chapter will explore the impact of moral distress on an individual's perspective and responsiveness toward self, others, and the world in the context of moral space,

© Springer Nature Switzerland AG 2020

K. Jones-Bonofiglio, *Health Care Ethics through the Lens of Moral Distress,*
The International Library of Bioethics 82,
https://doi.org/10.1007/978-3-030-56156-7_7

Fig. 7.1 Tensions of moral action

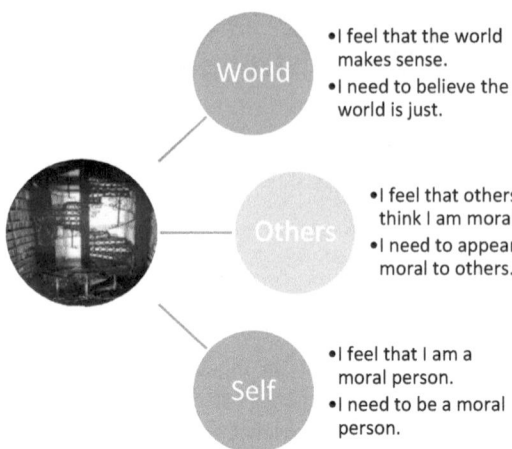

- I feel that the world makes sense.
- I need to believe the world is just.

- I feel that others think I am moral.
- I need to appear moral to others.

- I feel that I am a moral person.
- I need to be a moral person.

moral ambiguity, and moral oblivion. These circumstances can lead an individual to feel anxious and that they need to justify their integrity to others. Worse, an individual can externalize their experiences and take on bullying behaviours toward others; a 'get them, before they get you' approach. Or, an individual may simply choose to isolate themself to reduce their shame and angst and fully internalize their stress. Finally, impacts on an individual's view of the world can be devastating, shaking the very foundations of core beliefs about what is good and right in the universe. Such are the circumstances of moral injury. This three-tension approach to understanding morality situates moral distress as much more than just a negative response to an ethical situation.

7.2   Moral Space

North American nursing scholars, Elizabeth Peter and Joan Liaschenko (2004), note that it is the proximity of nurses to patients and their unique circumstances that creates a moral space where obligations can be recognized and nurses feel called to act. Although moral distress has been identified among many other health care professions, perhaps it is further potentiated for nurses by their proximity to suffering and the relational context of nursing practice.

Relationships themselves can also be considered a moral space, where both responsiveness and responsibility can be enacted (Bergum 2004). The nurse, as an individual, can be the moral space for a patient in the context of a therapeutic relationship (Quinn 1998). After all, it is in relationships that stories are created and shared. American physician and professor, Lewis Mehl-Madrona (2005), teaches that stories help us to organize, express, and live the meanings we make of our experiences.

'Holding space' can be one way to enact responsiveness and responsibility for someone's story or even for our own story. It means to bear witness, to support, and to offer comfort without judgement (Mehl-Madrona 2010). During this process, we maintain hope to achieve a better story one day and, in this solidarity with the other (or our self), we realize that we belong and are connected. This is the power of narrative.

7.3 Moral Ambiguity

What if the story doesn't make sense? Acceptance of uncertainty and a generous sense of humility are important approaches to take when complex ethical circumstances appear ambiguous. Moral ambiguity can be defined loosely as a lack of certainty (Merriam-Webster 2020a), but it can also be considered as an ability to be open to multiple interpretations. The latter description delineates moral ambiguity from moral uncertainty and moral disorientation. In fact, moral ambiguity rejects a strict dependence on moral rationalism (Van Roojen 2010) and leverages the synergies that can be found among the combined strengths of using both cognitive and affective approaches in decision making. Further, the concept of ethical pluralism denies ethics as a normative, prescriptive, paternalistic, expert reaction and allows for values and context sensitivity (Kovács 2010). This is important especially in cross-cultural contexts. However with this, there are still right and wrong choices. This kind of subjectivity can be, at best, uncomfortable and, at worst, set the stage for anger and rage reactions.

Health care has long had its roots in war and fighting battles. In fact those metaphors continue today as patients 'battle cancer' or 'fight for a cure.' Sometimes the battle happens where it should not. Bullying behaviours among health care providers are unfortunately not uncommon. In a study of over 9,000 direct care providers in long term care settings, bullying was linked to missed care for elderly clients (Hogh et al. 2018). In another study of 84 nurses employed in hospitals across the American midwest, bullying was linked to symptoms of psychological distress, such as anxiety and symptoms of post traumatic stress (Berry et al. 2016). Bullying is also referred to in the nursing literature as horizontal violence (Longo et al. 2016; Lewis-Pierre et al. 2019; Myers et al. 2016) and lateral violence (Chu and Evans 2016; Nemeth et al. 2017). Also, microaggressions (Freeman and Stewart 2018) against marginalized groups contribute to undermining quality patient care. Issues of bullying, violence, racism, prejudice, and discrimination that go unaddressed are major barriers to achieving safe places and spaces, health and wellbeing, and trust for and among health care providers and patients.

7.4 Moral Oblivion

American social psychologist, Stanley Milgram (1974), warns that any individual who is simply going about their work and minding their own business can be caught up in carrying out destructive and immoral actions. We have seen this to be true in such cases such as the Holocaust, the genocide in Rwanda, and the colonization of Indigenous people and land, as just three examples among many others. A place where morals have been abandoned can be called a state of moral oblivion or ethical apathy. The definition of 'oblivion' is a lack of awareness or a state of being forgotten or unknown (Merriam-Webster 2020b). This is the space where trauma can and does occur.

Post traumatic stress disorder (PTSD) is a psychiatric disorder that involves symptoms that can occur after a trauma and this results in psychosocial problems in living, loving, and doing that can last a lifetime (Olszewski and Varrasse 2005). In 1980, the diagnostic category of PTSD was accepted into the American Psychological Association's (APA's) Diagnostic and Statistics Manual (DSM-III) (Bonanno and Mancini 2012). This sparked tremendous research attention into trauma and its sequela. Today, the DSM-5 (APA 2013) notes the following symptom requirements for a diagnosis of PTSD:

- Indirect/direct exposure to trauma
- Trauma is persistently re-experienced (e.g., flashbacks, nightmares, etc.)
- Avoidance behaviours (related to potential triggers)
- Negative thoughts/feelings
- Arousal/reactivity behaviours
- Symptoms persist for over a month
- Symptoms interfere with everyday life
- Symptoms not attributed to other causes (e.g., medications, illness, etc.).

But, these are patients and not health care providers, right? Let's look into this a bit more. For example, an American study of 125 registered nurses who worked emergency medicine, critical care, or general nursing investigated for symptoms of PTSD (Kerasiotis and Motta 2004). What they found was high levels of anxiety likely due to numerous exposures to trauma and lack of control over these events. In another American study, this time with 173 pediatric acute care nurses, 21% of participants showed strong symptoms of PTSD along with comorbid mental health issues (e.g., anxiety, depression), and symptoms of burnout (Czaja et al. 2012). Further, a Canadian study of 51 health care providers from a single emergency room correlated PTSD symptoms with traumas related to interpersonal conflict and none sought help for their symptoms (Laposa et al. 2003). PTSD has also been found in psychiatric nurses, especially related to high risk for assaults and potential for patient violence (Jacobowitz 2013). Finally, an American study surveyed 63 emergency physician residents and found symptoms of PTSD (Mills and Mills 2005). Respondents noted that their symptoms increased over time and with subsequent exposures in the training program. Are the symptoms experienced by health care providers, as noted in these studies, all prodromal symptoms of PTSD or actual PTSD diagnoses? Perhaps there is more to consider.

7.5 Moral Injury

I first heard the term moral injury about four years ago. I instantly and instinctually connected with the concept, but felt that it would most likely be disrespectful to appear to 'borrow' it from its military context and attempt to apply it to nurses' experiences of moral distress. Now, physicians are using the term instead of burnout (Frezza 2019; Talbot and Dean 2018).

Moral injury is an interdisciplinary concept (Molendijk 2018) that describes intense guilt (may also include anger or shame) experienced by an individual who is in a state of existential disorientation (Gibbons-Neff 2015). The necessity lost in moral injury is one's sense of trust, of self, other, and/or the world due to a violation of deeply held values (Litz et al. 2009). See Table 7.1 to explore differences between symptoms of moral injury and PTSD.

I learned more about the concept of moral injury through the work of Jonathan Shay, an American clinical psychiatrist, who is famous for decades of work with military veterans. His definition of moral injury (Shay 2014) has three requisites:

1. A betrayal of what's right.
2. By a personal in authority.
3. In a high stakes situation.

Table 7.1 Moral injury and PTSD symptoms compared[a]

Moral injury	Shared symptoms	PTSD
Loss of trust etiology	–	Fear-based etiology
Guilt, shame, feeling unforgiveable, depression, demoralization	–	Hypervigilant, hyperaroused states
Spiritual crisis	–	–
Betrayal of what's right; person in authority; high stakes situation	–	–
Moral anguish, existential angst	–	–
–	Avoidance behaviours	–
–	Re-experiencing	–
–	Anxiety, anger, substance use	–

[a]Based on article by Buechner and Jinkerson (2016)

The good news is, Shay (2014) believes that there are ways to protect individuals from moral injury and these include efforts to build capacity for cohesion, leadership, and training. Further, the focus needs to be holistic and include both the person and their environment (Malabou and Miller 2012). We can recognize similar recommendations for these types of combined efforts in the context of moral distress. And what is the cost of not acting to address 'morally injurious experiences' (MIEs; Litz et al. 2009) or 'psychic wounding' (Malabou and Miller 2012) among health care providers? The cost is an erosion of trust and a shaken faith in humanity. It is the unnecessary loss (e.g., absenteeism, presenteeism, job turnover, retirement, addiction, suicide, etc.) of a valuable carer.

7.6   Trauma-Violence Informed Approaches

Lingchi is a form of torture known as 'death by a thousand cuts'. Many kinds of trauma, from simple to complex, can accumulate over time in peoples' lives and feel like a slow death. We may never truly known the burdens that others carry, unless they choose to share their story with us. Therefore, trauma and violence informed approaches are important for health care providers, in order to care for patients, families, communities, and for themselves as well. From an ethical perspective, trauma should be assumed as a universal precaution. Like donning a pair of gloves before a physical examination, health care providers should put on trauma-violence informed perspectives before engaging with others.

Trauma informed approaches are strength based, see individuals as survivors (versus victims), and accept that symptoms are adaptations rather than failure to cope (National Center on Domestic Violence, Trauma & Mental Health 2011). Trauma-Violence Informed Care (TVIC) acknowledges that people who have experienced trauma may feel unsafe, have had boundary violations, and have known abuses of power (EQUIP Health Care 2017; see Fig. 7.2).

7.7   Post-Traumatic Growth

American professor of clinical psychology, George Bonanno (2009), says that human resilience to trauma is not the exception to the rule- it is the rule! He has conducted extensive research with survivors of the 9/11 attacks in New York, as well as other ground breaking research in grief and trauma. He proposes that there are many pathways to resilience and has coined the term 'coping ugly' (Bonanno 2004). Resilience as the rule makes sense when you consider the fact that almost every person has had some type of a traumatic experience at some point in their lifetime (Norris and Slone 2013). Post traumatic growth is defined as positive change (e.g., appreciation, meaningful relationships, personal strength, priorities, spiritual life) after engaging with a crisis or trauma (Tedeschi and Calhoun 2004).

Fig. 7.2 Trauma-violence informed ways, inspired by EQUIP Health Care (2017)

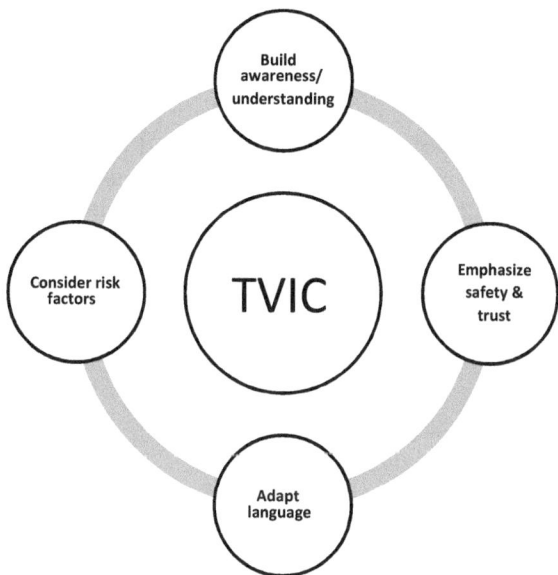

In an American study of 27 ICU nurses, resilience and PTSD were explored (Mealer et al. 2012). Nurses who were identified as highly resilient accessed spirituality, social supports, had a role model, and were optimistic. Nurses with a diagnosis of PTD lacked social supports, didn't have a role model, experienced disruptive thoughts and regret, and had lost their optimism. Four domains were revealed from this study: (1) self care; (2) social support network; (3) cognitive flexibility; and, (4) worldview.

7.8 Moral Community

Compassion for self and others makes room for resilience in our lives. I argue that compassion is one of the most valuable of the human virtues. Resilience can be considered to have two main factors. These include 'bounce back' and 'hardiness' (Jinpa 2015). The first is about capacity to navigate troubles and the second is about one's attitude toward adversity. These factors allow individuals to respond to adversity by feeling challenged (versus overwhelmed) and in control (rather than powerless).

Further, as noted in the studies above individuals need each other (e.g., social supports and role models). A strong moral community has high levels of both sociability and solidarity and supports ethical leadership (Sama and Shoaf 2008) and ethical practice. What might this look like?

In the book '*Leaders Eat Last: Why Some Teams Pull Together and Other's Don't*', key aspects of successful teams include having a circle of safety and belonging, sharing an identity and history, understanding a collective culture, feeling protected by leaders and colleagues, and feeling empowered in decisions and actions (Sinek 2014). Similarly, in the book '*The Culture Code: The Secrets of Highly Successful Groups*', the recipe for group cohesion involves safety, identity, trust, connection, interdependence, community, ethics, and vulnerability (Coyle 2018). The resources to address moral distress and to apply trauma and violence-informed perspectives to the experiences of fellow health care providers are at our fingertips. We are better together, but somehow we lack the time, the space, the energy, and the motivation to get together as a moral community.

> **Recap of Concept: Moral Community**
> ✓ Safety and belonging
> ✓ Identity and history
> ✓ Collective culture
> ✓ Protected and empowered
> ✓ Trust and interdependence.

7.9   Conclusion

There is a beautiful Japanese concept known as 'wabi-sabi'. It is a symbol of new beginnings and teaches that beauty is fleeting and imperfect. And so, our human condition is 'wabi-sabi' or flawed beauty. American nurse and theorist on caring, Jean Watson (2003) urges us to recall, "that it is our humanity that both wounds us and heals us, and those whom we serve; and in the end, it is only love that matters" (p. 99).

References

American Psychiatric Association. 2013. *Diagnostic and statistical manual of mental disorders*, 5th ed. Washington, DC: Author.

Bergum, V. 2004. Relational ethics and nursing. In *Toward a moral horizon: Nursing ethics for leadership and practice*, ed. J. Storch, P. Rodney, and R. Starzomski, 485–503. Toronto, Ontario, Canada: Pearson Education.

Berry, P.A., G.L. Gillespie, B.S. Fisher, D. Gormley, and J.T. Haynes. 2016. Psychological distress and workplace bullying among registered nurses. *OJIN: The Online Journal of Issues in Nursing* 21 (3): 4. https://doi.org/10.3912/OJIN.Vol21No03PPT41

Bonanno, G.A. 2004. Loss, trauma, and human resilience: Have we underestimated the human capacity to thrive after extremely aversive events? *American Psychologist* 59 (1): 20–28. https://doi.org/10.1037/0003-066X.59.1.20.

Bonanno, G. 2009. *The other side of sadness: What the new science of bereavement tells us about life after loss*. New York, NY: Basic Books.

Bonanno, G.A., and A.D. Mancini. 2012. Beyond resilience and PTSD: Mapping the heterogeneity of responses to potential trauma. *Psychological Trauma: Theory, Research, Practice, and Policy* 4 (1): 74–83. https://doi.org/10.1037/a0017829.

Buechner, B., and J. Jinkerson, J. 2016. Are moral injury and PTSD distinct syndromes? Conceptual differences and clinical implications. In *Veteran and family reintegration: Identity, healing, and reconcilliation,* eds. M. Guilarte and B. Buechner, 47–79. CreateSpace Independent Publishing Platform.

Chu, R.Z., and M.M. Evans. 2016. Lateral violence in nursing. *MedSurg Nursing* 25 (6): S4–S4.

Czaja, A.S., M. Moss, and M. Mealer. 2012. Symptoms of posttraumatic stress disorder among pediatric acute care nurses. *Journal of Pediatric Nursing* 27 (4): 357–365. https://doi.org/10.1016/j.pedn.2011.04.024.

EQUIP Health Care. 2017. *Trauma-and-Violence-Informed Care (TVIC): A tool for health & social service organizations and providers.* Vancouver, BC: Retrieved from https://equiphealthcare.ca/files/2019/12/TVIC-tool-ONTARIO-January-12-2018.pdf.

Freeman, L., and H. Stewart. 2018. Microaggressions in clinical medicine. *Kennedy Institute of Ethics Journal* 28 (4): 441–449. https://doi.org/10.1353/ken.2018.0024

Frezza, E. 2019. Moral injury: The pandemic for physicians. *Texas Medicine* 115 (3): 4–6.

Gibbons-Neff, T. 2015. Why distinguishing a moral injury from PTSD is important. *Stars and Stripes.* Retrieved from https://www.stripes.com

Haan, N. 1989. Coping with moral conflict as resiliency. In *The child in our times*, ed. T. Dugan and R. Coles, 23–42. New York, NY: Brunner-Mazel.

Hogh, A., M. Baernholdt, and T. Clausen. 2018. Impact of workplace bullying on missed nursing care and quality of care in the eldercare sector. *International Archives of Occupational and Environmental Health* 91 (8): 963–970. https://doi.org/10.1007/s00420-018-1337-0.

Jacobowitz, W. 2013. PTSD in psychiatric nurses and other mental health providers: a review of the literature. *Issues in Mental Health Nursing* 34 (11): 787–795. https://doi.org/10.3109/01612840.2013.824053.

Kerasiotis, B., and R.W. Motta. 2004. Assessment of PTSD symptoms in emergency room, intensive care unit, and general floor nurses. *International Journal of Emergency Mental Health* 6 (3): 121–133.

Kovács, J. 2010. The transformation of (bio) ethics expertise in a world of ethical pluralism. *Journal of Medical Ethics* 36 (12): 767–770. https://doi.org/10.1136/jme.2010.036319.

Laposa, J.M., L.E. Alden, L.M. Fullerton. 2003. Work stress and posttraumatic stress disorder in ED nurses/personnel (CE). *Journal of Emergency Nursing* 29 (1): 23–28. https://doi.org/10.1067/men.2003.7

Lewis-Pierre, L., D. Anglade, D. Saber, K.A. Gattamorta, and D. Piehl. 2019. Evaluating horizontal violence and bullying in the nursing workforce of an oncology academic medical center. *Journal of Nursing Management* 27 (5): 1005–1010. https://doi.org/10.1111/jonm.12763.

Litz, B.T., N. Stein, E. Delaney, L., Lebowitz, W.P. Nash, C. Silva, and S. Maguen. 2009. Moral injury and moral repair in war veterans: A preliminary model and intervention strategy. *Clinical Psychology Review* 29(8):695-706. https://doi.org/10.1016/j.cpr.2009.07.003

Longo, J., L. Cassidy, and R. Sherman. 2016. Charge nurses' experiences with horizontal violence: Implications for leadership development. *The Journal of Continuing Education in Nursing* 47 (11): 493–499. https://doi.org/10.3928/00220124-20161017-07.

Lützén, K., A. Cronqvist, A. Magnusson, and L. Andersson. 2003. Moral stress: Synthesis of a concept. *Nursing Ethics* 10 (3): 312–322. https://doi.org/10.1191/0969733003ne608oa.

Malabou, C., and S. Miller. 2012. *The new wounded: From neurosis to brain damage.* New York, NY: Fordham University Press.

Mealer, M., J. Jones, and M. Moss. 2012. A qualitative study of resilience and posttraumatic stress disorder in United States ICU nurses. *Intensive Care Medicine* 38 (9): 1445–1451. https://doi.org/10.1007/s00134-012-2600-6.

Mehl-Madrona, L. 2005. *Coyote wisdom: The power of story in healing.* Rochester, Vermont: Bear & Company.

Mehl-Madrona, L. 2010. *Healing the mind through the power of story: The promise of narrative psychiatry.* Rochester, Vermont: Bear & Company.

Milgram, S. 1974. *Obedience to authority: An experimental view.* New York, NY: Harper & Row.

Mills, L.D., and T.J. Mills. 2005. Symptoms of post-traumatic stress disorder among emergency medicine residents. *The Journal of Emergency Medicine* 28 (1): 1–4. https://doi.org/10.1016/j.jemermed.2004.05.009.

Mirriam-Webster. 2020a. *Moral ambiguity.* Retrieved from https://www.merriam-webster.com/dictionary/moral%20ambiguity

Mirriam-Webster. 2020b. *Oblivion.* Retrieved from https://www.merriam-webster.com/dictionary/oblivion

Molendijk, T. 2018. Toward an interdisciplinary conceptualization of moral injury: From unequivocal guilt and anger to moral conflict and disorientation. *New Ideas in Psychology* 51: 1–8. https://doi.org/10.1016/j.newideapsych.2018.04.006.

Myers, G., D. Côté-Arsenault, P. Worral, R. Rolland, D. Deppoliti, E. Duxbury, M. Stoecker, and K. Sellers. 2016. A cross-hospital exploration of nurses' experiences with horizontal violence. *Journal of Nursing Management* 24 (5): 624–633. https://doi.org/10.1111/jonm.12365.

National Center on Domestic Violence, Trauma & Mental Health. 2011. *Creating trauma informed services: Tipsheet series: Tips for creating a welcoming environment.* Retrieved from https://nationalcenterdvtraumamh.org/wp-content/uploads/2012/01/Tipsheet_Welcoming-Environment_NCDVTMH_Aug2011.pdf

Nemeth, L.S., K.M. Stanley, M.M. Martin, M. Mueller, D. Layne, and K.A. Wallston. 2017. Lateral violence in nursing survey: Instrument development and validation. *Healthcare* 5 (3): 33. https://doi.org/10.3390/healthcare5030033.

Norris, F.H., and L.B. Slone. 2013. Understanding research on epidemiology of trauma and PTSD. *PTSD Research Quarterly: Advancing Science and Promoting Understanding of Traumatic Stress* 24 (2–3): 1–13.

Olszewski, T.M., and J.F. Varrasse. 2005. The neurobiology of PTSD: Implications for nurses. *Journal of Psychosocial Nursing & Mental Health Services* 43 (6): 40.

Peter, E., and J. Liaschenko. 2004. Perils of proximity: A spatiotemporal analysis of moral distress and moral ambiguity. *Nursing Inquiry* 11 (4): 218–225. https://doi.or/10/1111/j.1440-1800.2004.00236.x

Quinn, J.F. 1998. Holding sacred space: The nurse as healing environment. *Essential Readings in Holistic Nursing* 6 (4): 84–93.

Sama, L.M., and V. Shoaf. 2008. Ethical leadership for the professions: Fostering a moral community. *Journal of Business Ethics* 78 (1–2): 39–46. https://doi.org/10.1007/s10551-006-9309-9.

Sethi, N., and J. Salinas. 2015. Autonomy and the "demanding encounter" in clinical neurology. *Neurology: Clinical Practice* 5 (3): 184–185. https://doi.org/10.1212/CPJ.0000466887.04124.b7.

Shay, J. 2014. Moral injury. *Psychoanalytic Psychology* 31 (2): 182. https://doi.org/10.1037/a0036090.

Sinek, S. 2014. *Leaders eat last: Why some teams pull together and others don't.* New York, NY: Penguin.

Talbot S.G., and W. Dean. 2018. Physicians aren't 'burning out.' They are suffering from moral injury. *STAT.* Retrieved from https://www.statnews.com/2018/07/26/physicians-not-burning-out-they-are-suffering-moral-injury/

Tedeschi, R.G., and L.G. Calhoun. 2004. Posttraumatic growth: Conceptual foundations and empirical evidence. *Psychological Inquiry* 15 (1): 1–18. https://doi.org/10.1207/s15327965pli1501_01.

Van Roojen, M. 2010. Moral rationalism and rational amoralism. *Ethics* 120 (3): 495–525. https://doi.org/10.1086/652302.

Watson, J. 2003. Loving and caring: Ethics of face and hand: An invitation to return to the heart and soul of nursing and our deep humanity. *Nursing Administration Quarterly* 27 (3): 197–202.

Chapter 8
Ethics, Leadership, and Compassion

Abstract Bioethics, medical ethics, nursing ethics, health care ethics, ethical dilemmas, and moral conflicts. Where is our common understanding of what ethics is and how it is applied in practice? A basic definition is that ethics are the principles that guide behaviour and/or conduct. In health care, ethics principles come from learned theories, professional codes of ethics and standards of practice, organizational policies and procedures, and personal values and beliefs. Ethical decisions are often influenced by laws and legislation with attention to the need to manage or mitigate risk and liability for more than one party. Rarely, if ever, do we enact ethics in a vacuum. In real life, ethics are about more than just knowing and following the rules. Ethics are lived and decision-making impacts in sometimes unanticipated and unintentional ways. Thus, ethics requires compassion (a willingness to suffer with) and leadership to support motivation to act.

Keywords Empathy · Ethics · Leadership · Compassion · Common humanity

8.1 Introduction

Ethics. Whose job is it anyway? Health care organizations may proudly display the fact that they have a bioethicist on staff, Monday to Friday between the hours of nine to five. A bioethicist can be an important resource for clinical ethics, research ethics, ethics education and training, policy and procedure review, and organizational issues. However, these same organizations may be reluctant to admit reduced capacity for ethics in practice on evenings, weekends, and holidays. Some health care providers may argue that they do not get paid for that; perhaps considering ethics as an add-on to basic, fundamental responsibilities that are formally required and accounted for. Some see their ethical responsibilities as knowing and following policy and procedures to the letter, without exceptions. Others claim a keen sense of professional ethics guided by personal convictions and virtues of character. In the absence of a sense of ownership of ethics in professional practice, health care providers can feel a sense of powerlessness and lack of control.

© Springer Nature Switzerland AG 2020   111
K. Jones-Bonofiglio, *Health Care Ethics through the Lens of Moral Distress*,
The International Library of Bioethics 82,
https://doi.org/10.1007/978-3-030-56156-7_8

We are taught that ethics are the principles that guide behaviour. If this is so, what informs the use of certain ethics principles (and not others) during a particular moral situation? In my work across almost two decades as a registered nurse, I have observed that ethical decisions in health care are driven largely by fear. Fear of the unknown, fear of past negative experiences, fear of punishment, fear of harm to others, and fear of making an already bad situation worse. Fear has a visceral power to create both physical and emotional responses. It is a primitive emotion that reduces the dynamic functions of our complex and creative human brains and privileges actions required for survival. Fear results in one of four limited options for response in the moment-fight, flight, freeze, or fawn (Hosier 2014). Also, living with the outcomes of poor ethical decision making, by self or by someone else, sets the stage for further fear and uncertainty. This can be a vicious, iterative cycle that builds additional fear and encourages disengagement from activities that will likely cause more of the same. This reality diminishes health care providers' capacity for enacting ethics in their everyday practice.

While a discussion of health care ethics often includes an emphasis on well-accepted western ethical principles, this chapter and this book will not. There are excellent academic resources that address various ethics principles thoroughly and thoughtfully (such as Beauchamp and Childress 2019; Hebert and Rosen 2019). Although complementary to discussions about ethics, normative ethics principles have not been given a place here as a source of moral refuge, in order to explore less familiar territory. This chapter aims to explore concepts such as ethics of care, virtue ethics, and relational ethics that more fully resonate with moral intelligence in nursing practice. This chapter argues for the strength of complementing a cognitive approach to ethical decision making with a receptive mode that includes acknowledging emotion and intuition (Noddings 2003).

8.2   Empathy

To begin, in perhaps an unexpected place, we will explore the notion of empathy among health care providers. Empathy is a human capacity, where nature and nurture each play a role. German Jewish philosopher Edith Stein wrote that acts of empathy teach us about who we are and allow a unique awareness of what we deem to be important in our lives (Stein 2012). Empathy is defined as being able to imagine and understand the experience(s) of another (Merriam-Webster 2020). Empathy is not necessarily an automatic function, rather it is a choice that we make (Brown 2018). The ability to be highly empathetic is an important pre-requisite for a health care provider to have.

English nursing scholar, Theresa Wiseman (1996), conducted a concept analysis on empathy and noted the following defining attributes: (1) nonjudgment; (2) understanding the feelings of other; (3) able to communicate understanding; and, (4) able to see the world as another sees it. Further to this work, Wiseman (2007) describes empathy as a continuum of development with four distinct forms of empathy. First she describes empathy as an incident and includes behavioural, cognitive, moral, and emotive components. Also, empathy is a way of knowing and empathy can be a process. Both of these perspectives on empathy contain aspects of relational knowledge. Finally, she describes empathy as a way of being; a form of embodied intelligence. It is not suprising to learn that Wiseman's research noted a lack of time as the number one barrier to expressing empathy to patients.

Nursing is known to be a caring profession that operates with attention and commitment to both empathy and connection (Fahrenwald et al. 2005). Empathy can act as a zero point of orientation or a frame of reference for ethics in practice. However, the notion of an 'ethics of empathy' as a guide for excellence in practice is highly contested for its vagueness and lack of direction (van Dijke et al. 2019). A second problem is that empathy can hurt. Empathy lights up parts of the human brain where pain and suffering is experienced. The brain does not differentiate between real suffering (of self) and imagined suffering (of the other). Suffering is suffering and empathetic responses can take their toll on a person.

In a Portuguese study, 280 nurses from public hospitals were surveyed to explore connections between empathy, self-compassion, and professional quality of life (Duarte et al. 2016). Their findings confirmed that high levels of empathy can put a health care provider at risk for compassion fatigue. They suggest that self-care and self-compassion be employed as interventions to reduce and mitigate negative outcomes.

In a French study of 24 residents (physicians) in general medicine rotations, researchers conducted interviews to explore ways that empathy might be connected to burnout (Picard et al. 2016). The participants suggested five different types of relationships, including: moderating factors (e.g. triggers), regulation (e.g., need for coping strategies), protection, psychological balancing (e.g., self care), and fatigue. They described empathy contributing to work satisfaction because it enabled them to help others, but at other times saw it as emotional labour.

Further, health care professions students are traditionally taught to establish a healthy professional distance from patients and their suffering in order to achieve a truly professional provider-patient relationship (versus a personal relationship) and to maintain the ability to be 'objective' in assessments, care, and decision making; to be a detached impartial observer (Parker 1990). The hidden curriculum is that health care providers are responsible to protect themselves. If moral harm occurs, they may face judgement that they should have been able to '*handle it*' or maybe '*you just don't have what it takes*'. Comments such as: '*suck it up, buttercup*' and '*toughen up, princess*' are common in health care among team members. I argue that health care providers should maintain a measure of their authentic self in their practice, despite this tradition. Many health care providers have developed skills (that they may not have learned in professional training) to appropriately employ therapeutic use of self and to maintain a unique caring identity and resilient caring practices; versus wearing a mask of professionalism or working on 'auto-pilot' (Jones-Bonofiglio and Vergidis 2020). It will be interesting to see if traditional perspectives change as new generations enter the health care workforce.

8.3 Virtues and Values

From the writings of Plato and Aristotle, virtue ethics is person-centred in that the character of the person making the ethical decision is of primary concern; an ethic of being versus an ethic of doing (Begley 2008). The application of virtue ethics to individuals and specifically to their profession (such as nursing) is not necessarily a well accepted concept. British philosopher Stephen Holland (2010) writes that to apply virtue ethics specifically to a profession, such as nursing, blurs the lines between personal and professional moral life and denies the abundant ethical resources that nurses have at their disposal.

Core nursing values (norms) and personal virtues of character guide nurses' moral behaviour worldwide in a variety of areas of ethically charged practice (Koskenvuori et al. 2019). It is noteworthy that character has been linked with clinical excellence in the nursing literature, because the ancient Greek understanding of the word *virtue* indicates excellence in relation to a skill or a function of character (Begley 2008). In a grounded theory study with Iranian nurses, personal characteristics of respect, commitment, responsibility, attentiveness, strong conscience, and philanthropy were linked to greater clinical competency in practice (Vanaki and Memarian 2009). Every virtue of a person's character can also be a vice. One view of this is that the flip-side of the gift is the curse; the light and the shadow (Brooks 2015). Another view is that virtues should not be considered as two-sided at all. For example, in *'Nicomachean Ethics'* Aristotle stressed the importance of the mean (Ameriks and Clarke 2000). For instance, the virtue of courage is neither cowardice nor foolhardiness (sort of too little and too much). See Table 8.1 for a list of examples of professional virtues (as defined by Irish nurse scholar Ann Marie Begley, 2008).

Western core nursing values include constructs such as dignity, social justice, integrity, altruism, and autonomy (Fahrenwald et al. 2005). These presumptions about shared values for professional nursing practice speak to the assumed requirements for development of good character and culturally accepted moral behaviour. These presumptions are normalized within society, the profession (sub-culture), and often internalized by individuals, perhaps without question or further reflection. Ethical congruency is the opportunity for personal values and professional work values to converge. This is an important juncture of being and doing that not all health care

Table 8.1 Professional nursing virtues[a]

Wisdom	Imagination	Courage
Veracity	Honesty	Justice
Benevolence	Patience	Kindness
Compassion	Generosity	Understanding
Integrity	Tolerance	Faithfulness
Diligence	Deliberation	Perseverance
Genuineness	Courtesy	Friendliness

[a]Inspired by Begley (2008)

providers experience. Ideally one's personal and professional values would closely align with one's roles and responsibilities as a health care provider.

For example, professional nursing values and ethical responsibilities as defined by the Canadian Nurses Association's code of ethics (CNA 2017) specifically include: (1) safe, compassionate, competent, ethical care; (2) health and well-being; (3) informed decision making; (4) dignity; (5) privacy and confidentiality; (6) justice; and, (7) accountability. Canadian nursing values are central to ethically responsible nursing actions such as providing, promoting, respecting, honouring, maintaining, and being. The American Nurses Association's code of ethics (ANA 2015) has nine provisions that are largely reflective of Canadian nursing values and includes additional content on a duty to care for self and a commitment to research and scholarship activities. Both codes identify that a nurse's foremost commitment is central to the *'patient'* (ANA 2015) or the *'person receiving care'* (CNA 2017).

8.4 Ethics and Leadership in Health Care

Scholars Marsha Fowler (professor of spirituality and ethics in the US) and Anne Gallagher (professor and nurse ethicist from the UK) did a study to compare ethical issues in nursing from 1919 to 2019. Over 100 years, the ethical issues in nursing seem to have changed very little. This is very interesting and perhaps even a little bit surprising. Yes, there are definitely new ethical issues now, such as those related to scientific advancements (e.g., stem cell research) and the use of technology (e.g., artificial intelligence algorithms for diagnosis). But some of the everyday ethical issues that nurses face have continued through the century seemingly unaddressed. From a shortage of nurses, to work overload, poor working conditions, high patient to nurse ratios, and fatigue (Fowler & Gallagher 2019), these century old ethical dilemmas still ring true today. How can this be?

While ethical issues themselves may have remained consistent, the moral intensity (Sama and Shoaf 2002) of professional practice in contemporary health care environments has only increased. The pace of practice and the sheer volume of new information and changes in technology present unique and unprecedented moral issues for health care providers.

Registered nurses in Canada are self-regulated by provincial nursing bodies who designate professional licensure and oversee investigations and disciplinary action. As such ethics in practice not only extends to patients, families, and communities in our care, but also to the necessary moral responsibility of reporting fellow nursing colleagues who may have breached standards of ethical practice. This duty to report makes a shared responsibility for ethics in practice a paradox. One the one hand, nurses need to work together toward high quality ethical care of patients. This moral work does not occur in isolation. On the other hand, nurses must report others who are not contributing positively to a strong moral culture and are perhaps participating in unethical acts or behaviours. From this vantage point, it would seem that nurses

must conduct the ethics of their practice independently of each other and from organizational pressures that influence the moral decisions that they make, yet must also
work interdependently.

Thus there is a growing need to build capacity for ethical leadership in the health
professions to validate individual care provider's moral values, sustain the moral
community of professionals, define moral culture in organizations and institutions,
support ethics in practice, and to uphold society's well-earned trust (Sama and Shoaf
2008).

In an American study of 100 nurse leaders, moral distress experiences in practice
were explored (Pavlish et al. 2016). Findings revealed multi-pronged experiences
of moral distress (e.g., affective, relational, physical, behavioral, and cognitive).
Key sources of moral distress were noted to be systemic factors, team conflict, and
diverse perspectives. Suggestions for strategies to address moral distress included:
(1) acknowledge it; (2) establish and maintain a culture of care; and, (3) increase
resilience (e.g., education, collaboration, support).

Leadership in health care organizations sets the tone for ethics in everyday practice. The moral ecology (or ethical climate) of an organization is represented by the
norms, beliefs, and behaviours that emerge in response to the needs of the moment,
and this shapes the people and actions within it (Brooks 2015). Absent moral leadership, from the top down, can lead to the development of a mode of survival where
people feel isolated and afraid to enact ethics in their practice (Sama and Shoaf 2008).

The ways that ethics are enacted in professional practice can vary. Is the rule
a guideline or an imperative (Noddings 2003)? If the rule is an imperative, then
an ethical decision likely has little wiggle room. However, solid ethical leadership
allows front line providers to navigate the ethics of rules that are set as guidelines and
trusts their knowledge and expertise. Figure 8.1 illustrates a standard 5-step process
for ethical decision making that is overlaid with a theory of five fundamental steps
for nurse leaders (Kouzes and Posner 2008).

8.5 Relational and Care Ethics

Relationships are considered to be a basic aspect of ontology, with caring relationships foundational to our ethicality (Noddings 2003). An ethics of caring grounds
nursing as a profession (Sawatzky et al. 2009; Wrubel and Benner 1989; Watson
1988). A relational ethic of care encompasses diverse and complex aspects of moral
experiences such as hope, powerlessness, courage, suffering, fear, and patience that
become etched into people's memories (Parker 1990).

8.6 Compassion as an Ethical Imperative

Compassion as a virtue should not be considered a means to an end, rather it should
be chosen for the sake of compassion itself (Begley 2008). A particularist approach

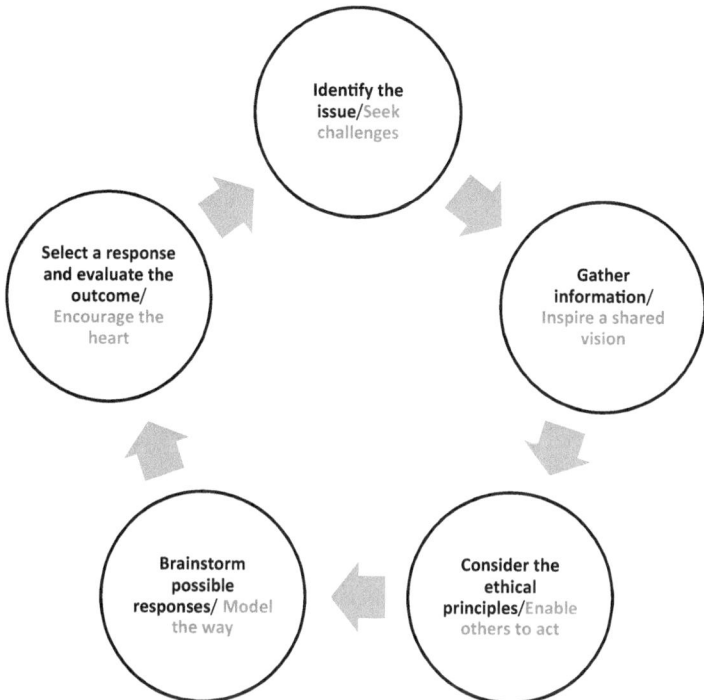

Fig. 8.1 Ethical decision making model with nursing leadership overlay

to ethics allows personal perspectives, such as values and virtues, to influence moral decision making and behaviour (Fahrenwald et al. 2005). This is an internal ethics, *written inside,* if you will, as opposed to an attempt to adhere to a code of ethics or rolled out only to obtain a particular desired outcome.

Compassion takes an individual experience and creates an opportunity for an intersubjective experience. Compassion can be broken down into the terms *passion* (to suffer) and *com* (with); to suffer with. Within this definition is the knowledge that the suffering is shared and that we are not alone. As such, compassion has the unique capacity to foster hope (Begley 2008).

Compassion is other focused and self-less action. It is not that you are not part of the work, it is that the focus is not on the ego or the self. Compassion is a choice and a commitment.

The philosophy of personalism affirms personal dignity and requires whole person service to others (Brooks 2015). Taking ethics in practice personally (or 're-personalizing' ethics) allows for the development of ethical knowledge through a deeply embodied and integrated process (Hartrick Doane 2002).

Ethical knowledge can be considered to be a type of 'phronesis' or practical wisdom, as described by Aristotle (Jenkins et al. 2019). But, this doesn't really explain how to *do* ethics. Roshi Joan Halifax (2009) teaches a metaphor about having a '*soft front and a strong back'* in order to do the work of caring. This imagery of needing to be strong and soft as required by the other can be combined with a meditation or a reflective practice. As a metaphor, the back is capable of lifting, able to flex and bend, carrying as needed on the shoulders, and pushing against if you need *'to get your back up'* (e.g., advocacy). The front is open and where the heart is located, capable of empathy and compassion for the other.

8.7   Compassion for Self

What about us- the health care providers? Perhaps health care providers make the worst patients. We are the ones who are used to giving advice and not usually the ones taking it. For example, a nurse smoking a cigarette on his break instead of eating a proper meal after teaching a patient about the importance of healthy eating and smoking cessation. *'Do as I say, not as I do.'* Similarly, most providers offer a level of comfort and caring to others that they may feel loathe to extend to themselves. Self-care often carries a stigma of selfishness and the fear that participating in self-kindness will be a slippery slope to narcissistic weakness. Author Nel Noddings (2003) writes of the ethical ideal of being our 'best self'' with the capacity to care and be cared for.

This is the paradox of compassion. In health care, providers should regularly be providing compassionate care to patients and families although many admit that there is limited time available for that to occur regularly. We cannot truly give what we don't have. But, once we have it for ourselves, the capacity to give it to others is actually much greater. All too often, compassion for others extends a far greater reach than compassion for self. The metaphor of putting on your own oxygen mask first before helping others seems to only receive merit in an aviation emergency situation.

Former Buddhist monk, translator to the Dalai Lama, and professor Thupten Jinpa (2015) writes that compassion is the common ground for all ethical teachings of each of the major religious and humanistic traditions and can be chosen as a basic human stance. Compassion is the foundation for responding to suffering in an ethically coherent manner. Opportunities for compassion exist in each moment and begin with attention to the present experience. This attention gives space for the possibility of empathy, understanding, and an empowered state of compassion (Jinpa 2015).

In his book, *'The Gift of Adversity'*, American medical doctor Norman Rosenthal (2013) tells a story about sage advice when teaching someone to swim; letting the swimmer hold the teacher (instead of the teacher holding the swimmer) to allow a sense of control and to allow the swimmer to let go when they are ready. This is compassion because it is other focused and the teacher is there for the student only for as long as they need it.

8.8 Common Humanity

Connection to others and the need to belong are universal human needs (Sinek 2017). Compassion allows us to recognize and relate to those needs and also see our own vulnerability reflected in others (Jinpa 2015). Often, fundamental needs for connection and belonging go unmet and this leaves individuals feeling isolated and unsupported in their practice and in their personal lives. Canadian Ojibway writer and scholar, Richard Wagamese (2016), writes of opportunities to be made more through being part of something beyond the self. These opportunities can be in the present and/or set the stage to be remembered in the future. This is the potential legacy of being a caring health care provider.

Being part of a community or a sub-culture, such as nursing, has the potential to allow for stories to be shared, meaning to be collectively interpreted, and for the tone to be set for a living moral foundation of practice. The language used to tell moral stories may change depending on who the audience is at the time. Language can be used to fully describe or fully distort reality. It has been said that to really know yourself and your nature, you need a community of others. Our own responsiveness to others teaches us something about our self.

Recap of Concept: Compassion
- ✓ Universal human need
- ✓ Basic human stance
- ✓ Response to suffering
- ✓ Part of something beyond the self
- ✓ Shared meaning and common ground
- ✓ Responsive connection
- ✓ Other focused.

8.9 Conclusion

Human lives have a moral core and the most important parts of our lives are the moments when we are faced with a moral choice and the desire to seek goodness and meaning (Brooks 2015). This necessitates health care providers to be willing to be present and attend to the circumstances of the moment. The start of a moral action begins with an initial moral perception (Sokol 2007) or awareness that an occasion for responsibility and moral choice is present. If you cannot recognize that an ethical issue has arisen, you are not likely to act on it. In a national study of the moral lives of hundreds of American college students, only 1/3 of the students could identify a moral problem that they had personally experienced (Smith et al. 2011). Health care providers must have concepts of authentic presence, self-awareness, and ethical decision making as part of their early and ongoing professional development.

Further, author Brooks (2015) cautions that we must recognize that a strong moral character is built upon crises and recovery, where any sense of mastery is wisely replaced with humility and an acceptance of lack of control. This translates into a message that perfection cannot be the goal. Ethics is complex, messy, and frustrating, even for those who claim to be skilled and experienced. A sense of moral courage must persevere despite unmet expectations, disappointments, and moral heartaches.

References

American Nurses Association (ANA). 2015. *Code of ethics with interpretative statements.* Silver Spring, MD: Author. Retrieved from https://nursing.rutgers.edu/wp-content/uploads/2019/06/ANA-Code-of-Ethics-for-Nurses.pdf

Ameriks, K., and D.M. Clarke. 2000. *Aristotle: Nicomachean ethics.* Cambridge, UK: Cambridge University Press.

Beauchamp, T.L., and J.F. Childress. 2019. *Principles of biomedical ethics,* 8th ed. Don Mills, Ontario: Oxford University Press.

Begley, A.M. 2008. Truth-telling, honesty and compassion: A virtue-based exploration of a dilemma in practice. *International Journal of Nursing Practice* 14 (5): 336–341. https://doi.org/10.1111/j.1440-172X.2008.00706.x.

Brooks, D. 2015. *The road to character.* New York, NY: Random House.

Brown, B. 2018. *Dare to lead.* New York, NY: Random House.

Canadian Nurses Association. 2017. *The code of ethics for registered nurses.* Ottawa, ON: Author. Retrieved from https://cna-aiic.ca/~/media/cna/page-content/pdf-en/code-of-ethics-2017-edition-secure-interactive.pdf?la=en

Duarte, J., J. Pinto-Gouveia, and B. Cruz. 2016. Relationships between nurses' empathy, self-compassion and dimensions of professional quality of life: A cross-sectional study. *International Journal of Nursing Studies* 60: 1–11. https://doi.org/10.1016/j.ijnurstu.2016.02.015.

Fahrenwald, N.L., S.D. Bassett, L. Tschetter, P.P Carson, L. White, and V.J. Winterboer. 2005. Teaching core nursing values. *Journal of Professional Nursing* 21 (1): 46–51. https://doi.org/1016/j.profnurse.2004.11.001

Fowler, M., and A. Gallagher. 2019. One hundred years of ethics in nursing: What's new? *Nursing Ethics* 26 (5): 1279–1281. https://doi.org/10.1177/0969733019862471.

Halifax, J. (2009). *Being with dying: Cultivating compassion and fearlessness in the presence of death.* Shambhala Publications.

Hartrick Doane, G. 2002. In the spirit of creativity: The learning and teaching of ethics in nursing. *Journal of Advanced Nursing* 39 (6): 521–528. https://doi.org/10.1046/j.1365-2648.2002.02320.x.

Hebert, P.C., and W. Rosen. 2019. *Doing right: A practical guide to ethics for medical trainees and residents,* 4th ed. Don Mills, Ontario: Oxford University Press.

Holland, S. 2010. Scepticism about the virtue ethics approach to nursing ethics. *Nursing Philosophy* 11 (3): 151–158. https://doi.org/10.1111/j.1466-769X.2010.00433.x.

Hosier, D. 2014. *Childhood trauma and its link to complex PTSD.* E-book (Kindle edition only): Childhoodtraumarecovery.com Publications.

Jenkins, K., E.A. Kinsella, and S. DeLuca. 2019. Perspectives on phronesis in professional nursing practice. *Nursing Philosophy* 2019 (1): e12231. https://doi.org/10.1111/nup.12231.

Jinpa, T. 2015. *A fearless heart: How the courage to be compassionate can transform our lives.* New York, N.: Hudson Street Press.

Jones-Bonofiglio, K., and D. Vergidis. 2020. I am a person 2.0. *Ethics, Medicine, and Public Health* 14:100422. https://doi.org/10.1016/j.jemep.2019.100422

Koskenvuori, J., O. Numminen, and R. Suhonen. 2019. Ethical climate in nursing environment: a scoping review. *Nursing Ethics* 26 (2): 327–345. https://doi.org/10.1177/0969733017712081.

Kouzes, J.M., and B.Z. Posner. 2008. *The leadership challenge* (4th ed.). San Francisco, California: Jossey-Bass.

Merriam-Webster. 2020. *Empathy.* Retrieved from https://www.merriam-webster.com/dictionary/empathy

Noddings, N. 2003. *Caring: A feminine approach to ethics and moral education,* 2nd ed. London, England: University of California Press Ltd.

Parker, R.S. 1990. Nurses' stories: The search for a relational ethic of care. *Advances in Nursing Science* 13 (1): 31–40.

Pavlish, C., K. Brown-Saltzman, L. So, and J. Wong. 2016. SUPPORT: an evidence-based model for leaders addressing moral distress. *JONA: The Journal of Nursing Administration* 46 (6): 313–320. https://doi.org/10.1097/NNA.0000000000000351

Picard, J., A. Catu-Pinault, E. Boujut, M. Botella, P. Jaury, and F. Zenasni. 2016. Burnout, empathy and their relationships: A qualitative study with residents in general medicine. *Psychology, Health & Medicine* 21 (3): 354–361. https://doi.org/10.1080/13548506.2015.1054407.

Rosenthal, N.E. 2013. *The gift of adversity: The unexpected benefits of life's difficulties, setbacks, and imperfections.* New York, NY: Penguin Books.

Sama, L.M., and V. Shoaf. 2002. Ethics on the web: Applying moral decision-making to the new media. *Journal of Business Ethics* 36 (1/2): 93–103.

Sama, L.M., and V. Shoaf. 2008. Ethical leadership for the professions: Fostering a moral community. *Journal of Business Ethics* 78 (1–2): 39–46. https://doi.org/10.1007/s10551-006-9309-9.

Sawatzky, J.A.V., C.L. Enns, T.J. Ashcroft, P.L. Davis, and B.N. Harder. 2009. Teaching excellence in nursing education: a caring framework. *Journal of Professional Nursing* 25 (5): 260–266. https://doi.org/10.1016/j.profnurs.2009.01.017.

Sinek, S. 2017. *Leaders eat last: Why some teams pull together and others don't.* New York, NY: Penguin Group.

Smith, C., K. Christoffersen, K.M. Christoffersen, H. Davidson, and P.S. Herzog. 2011. *Lost in transition: The dark side of emerging adulthood.* New York, NY: Oxford Press.

Sokol, D.K. 2007. Ethicist on the ward round. *BMJ: British Medical Journal* 335(Suppl S6). https://doi.org/10.1136/sbmj.0712467

Stein, W.J. 2012. *On the problem of empathy: The collected works of Edith Stein Sister Teresa Bendicta of the cross discalced carmelite volume three.* Springer Science & Business Media.

Vanaki, Z., and R. Memarian. 2009. Professional ethics: Beyond the clinical competency. *Journal of Professional Nursing* 25 (5): 285–291. https://doi.org/10.1016/j.profnurs.2009.01.009.

van Dijke, J., I. van Nistelrooij, P. Bos, and J. Duyndam. 2019. Care ethics: An ethics of empathy? *Nursing Ethics* 26 (5): 1282–1291. https://doi.org/10.1177/0969733018761172.

Wagamese, R. 2016. *Embers: One Ojibway's meditations.* Maderia Park, BC: Douglas & McIntrye.

Watson, J. 1988. *Human science and human care: A theory of nursing.* New York, NY: National League for Nursing.

Wiseman, T. 1996. A concept analysis of empathy. *Journal of Advanced Nursing* 23 (6): 1162–1167. https://doi.org/10.1046/j.1365-2648.1996.12213.x.

Wiseman, T. 2007. Toward a holistic conceptualization of empathy for nursing practice. *Advances in Nursing Science* 30 (3): E61–E72. https://doi.org/10.1097/01.ANS.0000286630.00011.e3.

Wrubel, J., and P. Benner. 1989. *The primacy of caring: Stress and coping in health and illness.* Menlo Park, CA: Addison-Wesley.

Chapter 9
Moral Courage

Abstract Moral confidence comes from a personal moral compass that can be trusted to help one navigate the, sometimes treacherous, ethical terrain of complex issues in contemporary health care practice. The ability to be guided by a sense of authentic meaning and purpose contributes to work satisfaction, professional fulfillment, and personal well-being. However, the ongoing recalibration and resilience that will be required throughout a career as a nurse or other valued health care provider may involve many unexpected lonely and painful existential searches in the darkness. This is not a deviation from a well worn path. This is the path and the essential parts of the journey. Exercising moral agency, setting healthy boundaries, holding realistic expectations, maintaining hope, and gaining ethical competence can offer important opportunities to work through moral distress experiences and build capacity for moral courage. And, it begins and ends with vulnerability.

Keywords Moral courage · Moral confidence · Moral agency · Ethical competence · Resilience

9.1 Introduction

Confidence can be loosely defined as a feeling of little doubt in someone, something, or yourself (Cambridge Dictionary 2020). Confidence comes from a triad of trust that allows an individual to cope with the conditions at hand, namely trust of self, trust of the environment, and trust of one's place in the world (Csikszentmihalyi 1990). Developing trust and confidence requires vulnerability, which is the precursor to courage (Brown 2017). And so, it seems pretty straight forward. Be vulnerable, then develop trust ($\times 3$), and finally establish a solid foundation of confidence. This is the simple recipe for moral courage and sustaining ethics in practice. So what is the problem?

There appears to be a chicken or the egg conundrum here. An ability and a willingness to be vulnerable has the potential to develop courage and character over time. However, it makes sense to expect that a measure of courage and strength of character are required to be willing to be vulnerable in the first place. And here's the crux of the

© Springer Nature Switzerland AG 2020
K. Jones-Bonofiglio, *Health Care Ethics through the Lens of Moral Distress*,
The International Library of Bioethics 82,
https://doi.org/10.1007/978-3-030-56156-7_9

problem- there are no guarantees. The risk is that vulnerability also has the potential to result in harm. Harm will not develop trust and a lack of confidence may occur and affect character over time. Although, a rare few may actually choose to "double down on vulnerability" (Brooks 2015 p. 96) even in the most difficult circumstances of life experiences, many individuals choose to err on the side of caution to avoid risk as much as possible.

Ethics often involves navigating a measure of risk. The amount of risk and what is at stake in the decision making process can vary depending on each unique ethical circumstance and for each individual. Some health care providers have had past positive experiences with ethics in practice and therefore have grown a sense of moral confidence. But what about those providers who have never allowed themselves to be vulnerable? Or worse, what about those who have had the foundations of their trust, confidence, and character badly shaken and now cannot garner the strength to find their way back from the ethical abyss? These difficult experiences may find health care providers unprepared or unsupported to endure or address them. Such experiences represent the deep roots of moral distress and require a better understanding of interconnected concepts and possible coping mechanisms.

In a Finnish concept analysis study, moral courage in nursing was found to have seven main attributes, which include: personal risk, perseverance and commitment, true presence, advocacy, honesty, responsibility, and moral integrity (Numminen et al. 2017). Further, these attributes are broken down into ways of 'being' courageous and ways of 'doing' or acting out courage. The authors note that moral courage has a central role in ethical nursing practice.

Those with moral courage are able to act on their values and principles (Comer and Vega 2011). Further, moral courage can also be described as a professional virtue of character in someone possessing values such as compassion, responsibility, respect, fairness, and honesty (Kidder 2006). However, English nurse scholar Ann Gallagher wisely suggests that individual moral courage may not be enough in all circumstances, and in fact, it is organizations that need to develop attributes of moral courage in order to support nurses and other health care providers to practice ethically (Gallagher 2011) (Fig. 9.1).

In an Iranian concept analysis of moral courage, Persian and Islamic contexts were explored to obtain a set of main attributes (Sadooghiasl et al. 2018). These attributes include right model orientation/acceptance, individual excellence, environment, and personal obstacles. Interestingly, these attributes are similar to other studies yet extend beyond the individual, indicating not individual or others/organization but rather both. The authors note that moral courage also depends on time and place and involves 'doing the right thing' despite known risks. This may result in self-actualization, where an individual achieves satisfaction through ethical risk taking and moral strength.

Fig. 9.1 Moral courage in nursing, inspired by Numminen et al. (2017)

9.2 Meaning and Purpose

Moral strength is about being securely rooted and willing to take on unforeseeable risks (Lindh et al. 2009). But where does this strength come from? Famous for his written work on finding meaning is professor, psychiatrist, and prolific author Viktor Frankl, survivor of World War II Nazi concentration camps. In his book, *'Man's Search for Meaning'*, he highlights that there are three key sources to finding meaning in life, namely significant work, loving others, and courageous suffering (Frankl 1985). Colloquially, this is known as having:

- *Something to do.*
- *Someone to love.*
- *Something to care about/hope for.*

Meaning can also be described as an overall assessment of one's sense of belonging (relationships) and ability to serve others (Seligman 2011). Therefore, finding meaning cannot occur in a vacuum. It includes doing, being, becoming, and connecting with others in sometimes uncomfortable ways that may include suffering (and often does). Meaning can be explained as our understanding of why; the rationale we use in order to make sense of the world we live in. However, for those individuals (including health care providers, emergency first responders [e.g., police, fire, medical] and military personnel) who repeatedly witness the extremes of human suffering, finding meaning can be more difficult than for someone not exposed to such traumatic realities.

The word *purpose* can be used as a noun or a verb; to have a reason (a purpose) or to intend (on purpose). Many individuals enter into a health care profession with the expressed purpose of wanting to make a difference. I know that I did. Having personally significant or purposeful work can be an important source of finding meaning. Further, intention or purposeful action can also be a notable factor in everyday ethical practice and ethical decision making. Generally speaking, the most ethical choice represents an individual's good intentions and is toward the purpose of the best possible outcome.

There is a careful and complementary balance that facilitates finding meaning and purpose, between what Hannah Arendt describes as a life of action (*vita activa;* external focus) and a life of reflection (*vita contemplativa*; internal focus) (Csikszent-mihalyi 1990). This is the space between doing and being; what I do may not define who I am but it contributes to who I become. Over time, I have come to reject the notion of balance or equilibrium in ethics. Demanding or striving to achieve balance can be a very unrealistic expectation that may leave individuals feeling that they have failed when in truth, they are seeking the impossible.

Today, I tend to support the notion of seeking harmony. The individual parts may not appear balanced, but they may balance each other as a whole. The sum of all the parts together can create a harmony, like a symphony. However, in the midst of a difficult ethical issue it may be very difficult to hear the music.

9.2.1 Mercy

At some point during the career of a nurse, moral distress will be experienced for the first time. Perhaps, it will first occur as a nursing student (Bordignon et al. 2019; Bordignon et al. 2018; Wojtowicz and Hagen 2014; Wojtowicz et al. 2014). It will likely happen again and again and moral residue (Webster and Baylis 2000; Epstein and Hamric 2009) will begin to stick and build up. Memory shards of difficult past ethical issues will become internally embedded, not visible to others but uniquely painful. In nursing school (or in undergraduate education for many health care professions), a protective forewarning of this may not be deeply explored or formally discussed. Likely, acquiring new fine motor skills and memorizing theory content will take precedence. With respect and to be fair, it may be impossible to truly and adequately forewarn future health care providers of the true rawness of the experiences they will have to face; the price they will be asked to pay related to moral distress experiences.

For example, recent studies have revealed a phenomenon that has long been kept quiet; nurses' substance use (Ross et al. 2018a). In a Canadian study, nurses shared their experiences of self-medicating their physical and emotional pain with substances within a culture of silence and harsh judgement (Ross et al. 2018b). Further, nurses often do not support each other in ways that they would be expected to support a patient with mental health and addictions issues. Common practices among nurses toward other nurses include othering, distancing, stereotyping, and concealing. As a collectively adopted and often internalized (stigmatizing moralistic) discourse, nurses with problematic substance use often must decide between the risks of remaining silent and the risks of seeking help. Nurses who do seek help report that workplaces and former colleagues regularly offered mistrust, hostility, suspicion, and contempt versus nonjudgmental support and genuine care.

In a Canadian study of experiences of voluntary alternative-to-discipline programs (ADPs), 12 nurses with substance involvement described being forced to actively engage with a standardized treatment model (one size fits all) that was not reflective of current, evidence informed, best practices for substance use disorders (Ross et al. 2019). Nurses in this study reported being forced to accept an 'expert' physician's mandatory treatment plan without input or questions, at the risk of losing their license to practice if they did not fully comply. They shared that they felt intimidated, threatened, fearful, coerced, and abused. Further the programming was not trauma-informed (e.g., such as promoting choice, empowerment), client-centred, or in a model of collaborative care (e.g., such as including the client's own care providers). The philosophical underpinnings of the harm reduction model were not applied (e.g., such as individualized care, meeting people where they are at in the treatment process). Set treatments required total submission and subordination, abstinence without any pharmacological assistance, and mandatory participation in a 12-step program. There was no consideration for the appropriate use of opioid agonist therapy (such as methadone or suboxone) as a first line treatment option or psychotropic medications. They also reported the mandatory use of specific private treatment services and labs (for blood work monitoring) that required out-of-pocket payment for two to five years (or longer) after beginning the ADP process. There are publicly funded services and labs available, but the nurses were told it would be inappropriate to access these. The 'voluntary' program denied any sense of autonomy and was in place of termination from their place of employment and risk of investigation into revocation of their nursing license.

Obviously, nurses' problematic substance use represents a significant safety issue, however ironically it is often issues of safety that underlie nurses' distress and subsequent substance involvement (Kunyk and Austin 2012). Within nursing, there is little room for repentance or the redemption of a *good nurse*, especially on the journey to recovery. There is often a high price to pay for being honest and courageously asking for help. Further, little attention is directed toward understanding how environmental factors and impacts may be contributing to nurses' substance use.

Also, the phenomenon of nurses' suicide is important to consider and bring to the forefront of discussions about the impact of practice realities on caring individuals. Substance use and suicidal behaviours are measures taken to manage pain and suffering. This pain can be physical (e.g., chronic back pain), emotional (e.g., anxiety, depression), or existential (e.g., loss of meaning and purpose). While pain is a universal experience, how much pain and suffering is tolerable or acceptable differs from person to person. There are many, sometimes compounding, variables to consider, such as intensity, frequency, past experiences, related losses, coping abilities, and sense of isolation. Negative coping strategies can become a tailspin of self-destruction that others may not know how to help with.

In a national study of Canadian medical students, high rates of mental distress, low mental quality of life, and six suicides (50% women) were reported over a ten year period (Zivanovic et al. 2018). Although this rate of suicide is lower than national rates, this issue needs to be addressed through prevention efforts as well as having protocols in place to guide action when a student dies by suicide.

For many health care professions students, the loss of a classmate can be their first real experience with grief and loss of someone their own age. In a Canadian study across four universities, nine undergraduate students shared their perspectives about losing a peer in the context of the culture of helping professions (Dorney 2016). Nine themes emerged, including: (1) emotional pain; (2) struggling with reality; (3) struggling with the void; (4) university/faculty responsiveness; (5) connecting; (6) bond of comfort with peers; (7) coping and support; (8) dwelling versus moving on; and, (9) grief as peer and nurse.

A study of deaths by suicide in the United Kingdom (UK), reveals that among carers the risk for suicide was double the national average and, among female health care professions, the risk was 24% higher (largely related to female nurse suicide) than the national average (Windsor-Shellard 2017). Sometimes suicide is related indirectly to the stress of the job and sometimes it is the actual job itself. In 2011, 24-year veteran critical care nurse, Kimberly Hiatt died by suicide seven months after a fatal medication error with a fragile baby, Kaia Z. (that may have contributed to the child's death) resulted in a professional investigation, and termination from the hospital (Aleccia 2011). This was nurse Hiatt's first serious medication error in almost a quarter century of professional nursing practice, she immediately reported the situation, and took full responsibility for accidentally administering ten times over the correct dosage of calcium to her little patient.

Addiction and suicide are the consequences of a culture of health care that leaves little room for forgiveness of self or others. Accurate statistics of suicides among health care providers are actually unknown (Davidson et al. 2018). In the absence of a sense of meaning and purpose, people may choose to pay the debt themselves. These circumstances occur within a culture of care that is currently limited. Confidence cannot grow under circumstances that punish vulnerability and unintentional failure. For nurses, the courage to come forward and to be held accountable is likely to be met with harsh judgment and penalty that may be unjust and life changing. Thus, there is a need for mercy among and for health care professionals. Mercy is compassionate forgiveness in an ethical context. Mercy is defined by having the power to condemn but choosing otherwise. Mercy is recognizing that the other could just as easily be ourselves. Mercy offers the opportunity to restore harmony, rebuild trust, and renew meaning and purpose.

9.3   Moral Agency

Over time, many health care providers wisely seek to adjust their personal sense of purpose from a novice's expectations *of saving the world* to more bite sized and realistic expectations about what they can reasonably offer to others in their daily practice. This is an important step in maturing and having a pragmatic perspective of one's roles and responsibilities. However, sometimes ongoing disappointments lead to the death of even our most reasonable expectations (Brown 2015). This can be source of courage drain, the beginning of hopelessness, or a call to reframe our perspective. Canadian nurse scholar, Elizabeth Peter, believes that opportunities for moral agency may not always be readily apparent (Peter 2018).

Perhaps a closer look at moral agency and how it is embodied (Rodney et al. 2013) is warranted. Who the moral agent is, or the character of the moral agent, is key to virtue ethics and dependent on striving to be a good person and/or living a good life (Racher 2007). Moral agency is the ability to make moral judgments and a willingness to be responsible for the outcomes (Peter 2011). Moral agency has been defined as having three components, namely,

(1) capacity to represent;
(2) sense of self; and,
(3) capable of authoritative judgement (Tirrell 1990).

A modern, refined definition of moral agency includes, (1) action resulting from insight; (2) willingness to dialogue; and, (3) readiness/ability to seek help (Milliken 2018). These concepts imply an opportunity for moral responsiveness and a moral responsibility to do so.

In a study of six nurses in western Canada, moral agency was described as an essential personal trait, duly influenced by organizational culture (Fortier and Malloy 2019). In this study, nurses felt constrained and disempowered by the politics of healthcare, workload, time, and resources/space (see Fig. 9.2). The politics of health care, in this study, refers to the commodification of care and the prioritizing of efficiency and cost effectiveness.

What about when moral agency has been constrained and the experiences of moral distress that may follow? English nurse scholar Aileen Walsh (2010) suggests that all of the emotions that arise during an ethical dilemma are not necessarily related to moral distress or even to moral issues at all, however, it is difficult to know the

Fig. 9.2 Constraints of moral agency, inspired by Fortier and Malloy (2019)

difference. She advocates for the need to reasonably limit one's sense of responsibility when appropriate advocacy has been carried out. Due to established boundaries of professional autonomy, nurses will continue to experience limits to their decision making and patient care outcome. In essence, these circumstances should be expected to come with the territory of nursing. Walsh questions if these circumstances should be considered a source of moral distress or, rather, as an acceptable limit to nurses' moral agency. In fact, some experiences of moral distress may actually be self-inflicted due to ineffective attempts at advocacy or professional authority (Berger et al. 2019). Such experiences represent an opportunity to proactively develop needed skills and confidence to manage future situations in a different way.

9.4 Healthy Boundaries

American feminist scholar bell hooks (2001) writes about individuals who lack the capacity to connect with others and how they do not take ownership for the pain that they cause. In essence she is highlighting that the opposite of moral distress is actually not moral satisfaction, it is psychopathology (e.g., lack of ownership for causing pain to others). Therefore, moral distress can be viewed as a healthy and normal response to being morally constrained. Perhaps it is the foundational context of relational practice in nursing that catalyzes the tendency for nurses to hold themselves accountable for the unnecessary pain and suffering of their patients in morally distressing situations. Where, then, should the boundaries of moral responsibility for nurses begin and where should they end?

For seasoned health care providers, their raison d'être may be to be a *good* care provider. Not a perfect nurse or flawless physician. This is a good and reasonable goal. A health boundary. American social work professor and scholar, Brene Brown, has a wise saying, "…when perfectionism is driving, shame is riding shotgun" (Brown 2015, p. 194). Shame is an isolating emotion and it can build to incapacitate action.

In a wonderful online posting by an Australian early childhood educator, Linda Newman (2018), she calls for redefining professional identity. Her message could easily be a call-to-action for nurses. She asserts that a profession is a shifting social construct and that in a profession (not an occupation) one must:

- Claim their power in their practice and make tough choices.
- Take ownership of knowledge and articulate it.
- Demonstrate behaviours grounded in values.
- Exercise autonomy.
- Take charge of professional learning.
- Collaborate.

9.5 Ethical Competence

The above points speak to developing professional competence. Competence and competencies are terms that are familiar to nursing. Nurses have a professional responsibility to be competent in their knowledge and skills for practice. In Canada, the College of Nurses of Ontario (CNO 2019) has newly revised their new graduate registered nurse entry-to-practice competencies and the framework includes nine roles required to be eligible for nursing registration in the province of Ontario (see Fig. 9.3). These requisite roles and responsibilities are also applicable to ethics in practice. However, American philosopher, Margaret Urban Walker (1993), warns that moral (ethical) competence cannot be reduced to an algorithm, much like the work of a carpenter cannot be reduced to that of a saw.

Within these various roles, nurses have moral responsibilities. In a Swedish study, the meaning of moral responsibility from the perspectives of 14 student nurses was explored (Lindh et al. 2007). Findings from three focus group sessions revealed an understanding of a 'relational way of being' with others, a sense of guidance from an individual's inner compass, and the desire to strive towards good in their nursing practice. Further, famous nursing scholar and theorist, Jean Watson, calls for an ethic of caring that considers context, continuity, and community (Watson 1990). What might that look like in a practical application of navigating ethical terrain? Canadian nurse scholar Gwen Hartrick, suggests a focus on building relational capacity to find harmony between the knowing and the doing of caring work. She suggests a responsive interpersonal approach that focuses on five capacities that may foster engagement (Hartrick 1997), which include:

1. Initiative, authenticity, and responsiveness
   –active concern through being, listening, and responding.
2. Mutuality and synchronicity
   –common goals; validate differences; rhythms of dialogue and silence.

Fig. 9.3 Examples of registered nurse competencies

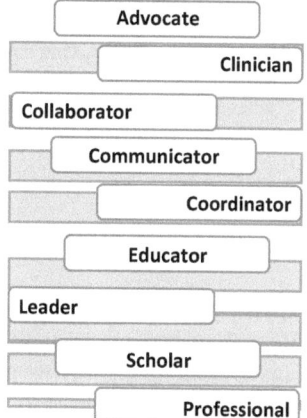

3. Honoring complexity and ambiguity
   –capacity to trust in uncertainty; curiosity; question, clarify, and connect.
4. Intentionality in relating
   –clear values; clarify meaning; discover choice and power.
5. Re-imagining
   –living the question; transforming meaning and experiences.

The practical aspects of ethical competence can also be called 'reckoning'. Reckoning means to calculate where you are now, where you came from, how you got to here, and next steps forward (Brown 2015). American nurse scholar, Alvita Nathaniel (2006), uses the term 'moral reckoning' to describe her theory of a three stage process: (1) stage of ease; (2) values conflict; (3) resolution (give up or take a stand). After this process, the nurse is challenged to share their story, examine the experience, and to fully participate in moral decision making.

In order to do this, some scholars advocate that it is through sharing stories of ethics in practice that nurses become moral agents (Skott 2003). Storytelling in nursing is both a time-honoured tradition and a therapeutic need for our dialogic community of practice (Maeve 1994). "Stories tell us what we are capable of, and so they tell us who we might be" (Tirrell 1990 p. 117). There are two stories from Greek mythology that I will share with you now in my words and in my own way. You can take the lessons that you need to from these stories.

> Let's begin with Icarus who was the son of a master craftsman. In fact, his father created the world's first labyrinth. Icarus and his father lived on an island. When there was a need to escape, his father constructed beautiful wings out of feathers and wax. Before taking flight, the father warned his son first, about too much confidence and second, about becoming complacent. He said, '*My son, do not fly to high or too low. Too high and the sun will melt the wax and too low, the sea will dampen the feathers.*' At first Icarus recalled his father's advice, but soon he forgot the knowledge he had been gifted. Eventually, Icarus rose higher and higher in the sky until the wax melted and he fell into the sea.

> The second story is about Chiron (pronounced K-eye-ron). Chiron was a Centaur. Yes, he was half man and half horse. I don't know if you've heard this, but centaurs don't have the best reputation. However, Chiron was different. He was a great healer, an accomplished astrologer, and a respected oracle (prophet or guide). He had great knowledge and many skills in medicine. He had a calm and patient manner. He was intelligent and kind. He only ever cared for others. One day Chiron was accidentally struck by a poisoned arrow. Over the course of many days and many nights, he tried to use his medical knowledge to treat the wound. He used all of his skills and resources. He tried desperately to heal himself, but he could not. Eventually, his wise and gentle soul was lifted toward the stars in the sky.

9.6 Recalibration

Swiss psychologist, Carl Jung (2014) suggests that humans are born whole, with a life task to become an individual (true self; individuation). If we lose touch with ourselves along the way we get sick. Therefore there is a need to nourish the self and to address grief and loss; to attend to moral distress experiences and moral residue. Self-care is a term that may not be welcomed by health care providers. I invite you to call it what you need to: recalibration, renovation, restructuring, rehabilitation, recovery, reparation, restoration. Ask yourself what is needed to maintain your sense of trust and vulnerability? How can you best support your capacity for courage and character?

9.7 Resilience

Resilience is required of all humans. For health care providers, resilience is a means of remaining whole in a broken system and retaining essential humanity in your personal and professional practices. Another term for resilience is existential hardiness (Fortier and Malloy 2019), which seems to be best supported by nurse's practical wisdom, experience, and peer support. In a validated measure of resilience (Connor and Davidson 2003), five factors were highlighted: (1) personal competence; (2) spiritual influences; (3) acceptance of change; (4) control; and, (5) secure relationships. There are many ways that individuals can build capacity for resilience in little ways on a daily basis across these areas.

My favourite metaphor for resilience is the Japanese art of 'kintsugi'. This is the repair of broken pottery with gold to highlight the brokenness instead of trying to hide the flaws. This process actually gives the dish more value. It highlights a proud history of needing to be mended (vulnerability) and honestly acknowledges imperfections (trust). Resilience brings us full circle back to confidence and courage. Difficult experiences are the fires that forge hard steel, the gritty bits that create pearls, and the intense pressures that make diamonds.

> **Recap of Concept: Resilence**
> ✓ Personal values
> ✓ Meaning and purpose
> ✓ Acceptance
> ✓ Control (where it can be found)
> ✓ Connections to others.

9.8 Conclusion

The evolving narrative of nursing might be best described by connecting the meaning of human living, healing, and wholeness (Spenceley 2004). Among health care providers and their many stories about daily encounters with others, there is a multi-coloured mosaic of profound and mundane moments of humanity. Society sees health care providers generally as contributing to a calling, a vocation, a respected profession, a fulfilling career, or a dream job. The choice to do caring work and to be a health care provider was hopefully chosen purposefully for the meaning of what it would contribute to others. That sense of meaning and purpose may or may not always be retained.

I have heard it said before that most people do not roll out of bed in the morning with the explicit intention to have the worst effect on the largest amount of people possible. But, we have all met people who appears to do just that. In general, we try to do our best in the time and with the resources that we are given. The challenge is to build and sustain moral confidence under those limited circumstances. To be willing to keep trying, trusting, and to be open to vulnerability. Courage, from the old French and Latin words for '*heart*' (Merriam-Webster 2020), is truly heart work among caring and engaged health care professionals. And this cannot be accomplished alone. The African philosophy of '*ubuntu*', to be and become with and through other persons, reminds us that it is collectivism and not isolated individualism that helps us to find not only our ethics but also ourselves (Haegert 2000).

References

Aleccia, J. 2011. Nurse's suicide highlights twin tragedies of medical errors. *MSNBC.com.* Retrieved from https://www.cmpa-acpm.ca/static-assets/pdf/education-and-events/workshops/theatre-art-twin-tragedies-of-medical-error.pdf

Berger, J.T., A.B. Hamric, and E. Epstein. 2019. Self-inflicted moral distress: Opportunity for a fuller exercise of professionalism. *Journal of Clinical Ethics* 30 (4): 314–317.

Bordignon, S.S., V.L. Lunardi, E.L.D. Barlem, G.D.L. Dalmolin, R.S. da Silveira, F.R.S. Ramos, and J.G.T. Barlem. 2019. Moral distress in undergraduate nursing students. *Nursing Ethics* 26 (7–8): 325–2339. https://doi.org/10.1177/0969733018814902

Bordignon, S.S., V.L. Lunardi, E.L. Barlem, R.S.D. Silveira, F.R. Ramos, G.D.L. Dalmolin, and J.G.T. Barlem. 2018. Nursing students facing moral distress: strategies of resistance. *Revista Brasileira de Enfermagem* 71: 1663–1670. https://doi.org/10.1590/0034-7167-2017-0072

Brooks, D. 2015. *The road to character.* New York, NY: Random House.

Brown, B. 2015. *Rising strong: The reckoning. The rumble. The revolution.* New York, N.Y.: Spiegel and Grau.

Brown, B. 2017. *Braving the wilderness: The quest for true belonging and the courage to stand alone.* New York, NY: Random House.

Cambridge Dictionary. 2020. *Confidence.* Retrieved from https://dictionary.cambridge.org/dictionary/english/confidence

College of Nurses of Ontario (CNO). 2019. *Entry to practice competencies for registered nurses.* Toronto, ON: Author. Retrieved from https://cno.org/globalassets/docs/reg/41037-entry-to-practice-competencies-2020.pdf

Comer, D.R., and G. Vega. 2011. The relationship between the personal ethical threshold and workplace spirituality. *Journal of Management, Spirituality & Religion* 8 (1): 23–40. https://doi.org/10.1080/14766086.2011.552255.

Csikszentmihalyi, M. 1990. *Flow: The psychology of optimal experience.* New York, NY: Harper Perennial.

Davidson, J., J. Mendis, A.R. Stuck, G. DeMichele, and S. Zisook. 2018. Nurse suicide: Breaking the silence. *NAM Perspectives.* Retrieved from https://pdfs.semanticscholar.org/7e0f/be794405c8cd8290506752b45fc01310ce87.pdf

Dorney, P. 2016. The empty desk: The sudden death of a nursing classmate. *OMEGA-Journal of Death and Dying* 74 (2): 164–192. https://doi.org/10.1177/0030222815598688.

Epstein, E.G., and A.B. Hamric. 2009. Moral distress, moral residue, and the crescendo effect. *Journal of Clinical Ethics* 20: 4.

Fortier, E., and D. Malloy. 2019. Moral agency, bureaucracy & nurses: A qualitative study. *The Canadian Society for Study of Practical Ethics*/Société Canadienne pour L'étude de L'éthique Appliquée (SCÉÉA) 3 (1). Retrieved from https://scholar.uwindsor.ca/csspe/vol3/1/1/

Frankl, V.E. 1985. *Man's search for meaning*. Boston, Massachusetts: Beacon Press.

Gallagher, A. 2011. Moral distress and moral courage in everyday nursing practice. *Online Journal of Issues in Nursing* 16 (2). https://doi.org/10.3912/OJIN.Vol16No02PPT03

Haegert, S. 2000. An African ethic for nursing? *Nursing Ethics* 7 (6): 492–502. https://doi.org/10.1177/096973300000700605.

Hartrick, G. 1997. Relational capacity: The foundation for interpersonal nursing practice. *Journal of Advanced Nursing* 26 (3): 523–528. https://doi.org/10.1046/j.1365-2648.1997.t01-12-00999.x.

hooks, b. 2001. *All about love: New visions*. New York, NY: HarperCollins Publishers.

Jung, C.G. 2014. *The archetypes and the collective unconscious*. New York, NY: Routledge.

Kidder, R. 2006. *Moral courage*. New York, NY: HarperCollins Publishers.

Kunyk, D., and W. Austin. 2012. Nursing under the influence: A relational ethics perspective. *Nursing Ethics* 19 (3): 380–389. https://doi.org/10.1177/0969733011406767.

Lindh, I., E. Severinsson, and A. Berg. 2007. Moral responsibility: A relational way of being. *Nursing Ethics* 14 (2): 129–140. https://doi.org/10.1177/0969733007073693

Lindh, I.B., E. Severinsson, and A. Berg. 2009. Nurses' moral strength: a hermeneutic inquiry in nursing practice. *Journal of Advanced Nursing* 65 (9): 1882–1890. https://doi.org/10.1111/j.1365-2648.2009.05047.x.

Maeve, M.K. 1994. The carrier bag theory of nursing practice. *Advances in Nursing Science* 16 (4): 9–22. https://doi.org/10.1097/00012272-199406000-00004.

Merriam-Webster. 2020. *Courage*. Retrieved from https://www.merriam-webster.com/dictionary/courage

Milliken, A. 2018. Refining moral agency: Insights from moral psychology and moral philosophy. *Nursing Philosophy* 19 (1): e12185. https://doi.org/10.1111/nup.12185.

Nathaniel, A.K. 2006. Moral reckoning in nursing. *Western Journal of Nursing Research* 28 (4): 419–438. https://doi.org/10.1177/0193945905284727.

Newman, L. 2018. Professionalism, ethics, and the meanderings of a secret nanna. Retrieved from https://publications.ieu.asn.au/2018-october-bedrock/articles1/professionalism-ethics-and-meanderings-secret-nanna/

Numminen, O., H. Repo, and H. Leino-Kilpi. 2017. Moral courage in nursing: A concept analysis. *Nursing Ethics* 24 (8): 878–891. https://doi.org/10.1177/0969733016634155.

Peter, E. 2011. Fostering social justice: The possibilities of a socially connected model of moral agency. *The Canadian Journal of Nursing Research/Revue Canadienne de Recherche en Sciences Infirmieres* 43 (2): 11–17.

Peter, E. 2018. Overview and summary: Ethics in healthcare: Nurses respond. *OJIN: The Online Journal of Issues in Nursing* 23 (1). https://doi.org/10.3912/OJIN.Vol23No01ManOS

Racher, F.E. 2007. The evolution of ethics for community practice. *Journal of Community Health Nursing* 24 (1): 65–76. https://doi.org/10.1080/07370010709336586.

Rodney, P., S. Kadyschuk, J. Liaschenko, H. Brown, L. Musto, and N. Snyder. 2013. Moral agency: Relational connections and support. In *Toward a moral horizon: Nursing ethics for leadership and practice*, 2nd ed., ed. J.L. Storch, P. Rodney, and R. Starzomski, 160–187. Toronto, ON: Pearson Prentice Hall.

Ross, C.A., N.S. Berry, V. Smye, and E.M. Goldner. 2018a. A critical review of knowledge on nurses with problematic substance use: The need to move from individual blame to awareness of structural factors. *Nursing Inquiry* 25 (2): e12215. https://doi.org/10.1111/nin.12215.

Ross, C.A., S.L. Jakubec, N.S. Berry, and V. Smye. 2018b. "A two glass of wine shift": Dominant discourses and the social organization of nurses' substance use. *Global Qualitative Nursing Research* 5: 1–12. https://doi.org/10.1177/2333393618810655.

Ross, C.A., S.L. Jakubec, N.S. Berry, and V. Smye. 2019. The business of managing nurses' substance-use problems. *Nursing Inquiry* 27 (1): e12324. https://doi.org/10.1111/nin.12324.

Sadooghiasl, A., S. Parvizy, and A. Ebadi. 2018. Concept analysis of moral courage in nursing: A hybrid model. *Nursing Ethics* 25 (1): 6–19. https://doi.org/10.1177/0969733016638146.

Seligman, M.E.P. 2011. *Flourish: A visionary new understanding of happiness and well-being.* New York, NY: Free Press.

Skott, C. 2003. Storied ethics: Conversations in nursing care. *Nursing Ethics* 10 (4): 368–376. https://doi.org/10.1191/0969733003ne619oa.

Spenceley, S.M. 2004. Out of fertile muck: The evolving narrative of nursing. *Nursing Philosophy* 5 (3): 201–207. https://doi.org/10.1111/j.1466-769X.2004.00187.x.

Tirrell, L. 1990. Storytelling and moral agency. *The Journal of Aesthetics and Art Criticism* 48 (2): 115–126. https://doi.org/10.2307/430901.

Walker, M.U. 1993. Keeping moral space open: New images of ethics consulting. *Hastings Center Report* 23 (2): 33–40. https://doi.org/10.2307/3562818.

Walsh, A. 2010. Pulling the heartstrings, arguing the case: A narrative response to the issue of moral agency in moral distress. *Journal of Medical Ethics* 36 (12): 746–749. https://doi.org/10.1136/jme.2010.036079.

Watson, J. 1990. Response to: Reconceptualizing nursing ethics. *Scholarly Inquiry for Nursing Practice* 4 (3): 219–221.

Webster, G.C., and F. Baylis. 2000. Moral residue. In *Margin of error: The ethics of mistakes in the practice of medicine*, ed. S.B. Rubin and L. Zoloth, 217–230. Haggerstown: University Publishing Group.

Windsor-Shellard, B. 2017. *Suicide by occupation, England: 2011 to 2015.* London, England: Office for National Statistics

Wojtowicz, B., and B. Hagen. 2014. A guest in the house: Nursing instructors' experiences of the moral distress felt by students during inpatient psychiatric clinical rotations. *International Journal of Nursing Education Scholarship* 11 (1): 121–128. https://doi.org/10.1515/ijnes-2013-0086.

Wojtowicz, B., B. Hagen, and C. Van Daalen-Smith. 2014. No place to turn: Nursing students' experiences of moral distress in mental health settings. *International Journal of Mental Health Nursing* 23 (3): 257–264.

Zivanovic, R., J. McMillan, C. Lovato, and C. Roston. 2018. Death by suicide among Canadian medical students: A national survey-based study. https://doi.org/10.1177/0706743717746663.

Chapter 10
Navigating Moral Distress

Abstract Moral distress; it can break you or build you. It is a complex, dynamic concept with many interconnected layers. However, it always begins with an individual experience. A moment. A sense that what is happening should not be (or what is not happening should be). There is a desire to act differently, either in the moment or upon reflection on that moment. For some, this is an embodied experience, carried within; a burden and a consequence of caring for, with, and about. In nursing, moral distress is also a collective experience. Although nurses rarely talk about it, there is a shared understanding that we have all been to that familiar terrain. We can relate to the smell, touch, taste, sound, and sight of it. When we listen to stories that are shared, often we can hear our own experiences echoed in the words of another health care provider. Discussed here are ways to navigate moral distress and how nurses and other care providers might consider moral distress as a valuable resource and a catalyst for positive changes in their ethical and moral endeavours.

Keywords Wellbeing · Education · Collaborative practice · Knowledge Power

10.1 Introduction

This book encourages a healthy respect for health care providers' moral distress experiences. I argue that moral distress can be a lens to view and better understand ethics in health care environments and professional practice. And there is still more work to do. I am not alone in suggesting that the concept of moral distress needs to be reworked and perhaps broadened (Campbell et al. 2016), although not restructured so widely that it loses all meaning (McCarthy and Deady 2008). American nurse ethicist and scholar, Cynda Rushton, also calls for a re-orientation of our understanding of moral distress with a focus on empowering clinicians to create positive change (Carse and Rushton 2017).

Now, let's look at what can be done about moral distress in practice and the significance of acting on it. Respect, in the original meaning of the word, means to observe (*spect*) again (*re*). So, to re-spect moral distress is to consider a new view of a concept that has been studied for four decades and to continue to explore

© Springer Nature Switzerland AG 2020   137
K. Jones-Bonofiglio, *Health Care Ethics through the Lens of Moral Distress*,
The International Library of Bioethics 82,
https://doi.org/10.1007/978-3-030-56156-7_10

opportunities for moral responsiveness (Lovecky 1997) to everyday ethical issues in practice. Until 2012, there were only three key intervention studies that attempted to directly address moral distress (Beumer 2008 [workshop]; Rogers et al. 2008 [educational sessions]; and Kälvemark Sporrong et al. 2007 [education + ethics rounds]) and all involved an ethics education approach.

Ethics can be widely described as a moral compass that aids navigation of the often difficult terrain of everyday practice in contemporary health care environments. Although, many health care providers still view ethics as narrowly pertaining to the big moral issues of life and death, technology, and science. Little described is ethics as professional radar and a learned capacity to detect morally relevant nuances up close and at a distance. The ability to anticipate ethical challenges can give health care providers valuable time to problem solve in order to mitigate or even prevent negative outcomes related to ethical issues. Interventions and training that build capacity for a more holistic view of ethics in practice and among multidisciplinary health care providers strengthens the health care team, the organization, the health care system, as well as individuals. The time, space, knowledge, and resources to identify and address ethical issues, a common language to describe these experiences, and a variety of skills to promote resilience all contribute to perceiving moral distress as a guidepost rather than a personal stumbling block or an avoidable risk to organizational integrity.

10.2   Health and Wellbeing in Health Care

It has long been acknowledged that generally 'health care' is a misnomer and in fact what the system really provides is essentially illness care. It is time to focus attention on the added layer of illness that often goes undetected and is comprised of high levels of unresolved (and perhaps unacknowledged) moral distress among health care providers. Like a slow growing cancer, the symptoms can be easily attributed to other factors or blamed on individual weaknesses until it is too late.

Beyond just illness care is the promotion and protection of health and wellbeing; a valuable upstream approach. This is important for patients, families, and communities as well as health care providers. In fact, I argue that the promotion and protection of health and wellbeing is a win–win on many levels as it also contributes to professional satisfaction. Potential for thriving, flourishing, meaning-making, purpose-finding, and joy are part of what brings people into caring professions in the first place and also what keeps them there despite the inherent hardships and challenges. These types of positive experiences are the backbone of resiliency and hope. Whether individuals come to work in health care because they feel chosen (vocation), make the choice (career), or are seeking a good paycheck (job), the reasons for coming in may actually be connected to the reasons for leaving.

A disconnect may occur between dearly held values and beliefs and repeated stark experiences of reality that impact one's ability to be healthy and well as an employee (and as a human) in a health care environment. It may be believing that you, personally, are not cut out for this important work; believing that you cannot make a positive contribution despite your best efforts; or, that you don't get paid enough to put up with such difficulties. This is about more than just being unhappy at work; it can make health care providers sick and tired (of being sick and tired).

American psychologist Martin Seligman (2011) has a popular theory of well-being and describes five key elements with the mnemonic of PERMA (P—positive emotion; E—engagement; R—positive relationships; M—meaning; and, A—accomplishment). This is a holistic theory that encompasses emotional, social, relational, spiritual, and physical components of achieving wellbeing. His work within this framework has led to research with the United States Army and the development of a tool to measure and track psychosocial fitness. Known as the Global Assessment Tool (GAT), a questionnaire is used to self-assess domains of social, emotional, family, and spiritual fitness in terms of both strengths and weaknesses (Seligman 2011). With nursing's historic roots running deeply into battlefields and religious orders respectively, perhaps it is time to bravely turn to both tactical and practical strategies to promote the health and wellbeing of contemporary health care providers?

10.3 Post-Secondary Ethics Education

To begin, the training and professional socialization of the next generation of health care providers is a good place to start. Nursing, as a prime example, is referred to as being both a science and an art. Nursing theory and academic literature often refer to the need to consider not only knowledge and skills, but also attitudes and beliefs. However, the course content that is often privileged, prioritized, and rewarded (in terms of required readings, lectures, assignments, and grades) in pre-licensure education say something about what is valued in a profession. In health care professions education, usually it is the 'hard science' content that takes the lead. In nursing specifically, it is often the medical-surgical, acute care, bio-medical model content that is the core of most western-style nursing curricula. Some limited room is likely to be made for specialty areas (e.g., obstetric, paediatric, geriatric, mental heath, community nursing) and considerations for unique aspects of health and (mostly) illness across the lifespan. And yes, caring. We teach a little of that too. Everything else is an extra. Not to diminish the value of long standing sacred cows in health care professions curricula, but there is other important content that must find a place and this will require innovative pedagogies in order to do this.

The purposeful inclusion of evidence-informed content on ethical decision making and moral distress experiences must become a core aspect of post-secondary education for today's health care professions students. However, the drive to achieve a competency-based curriculum, with an ever increasing amount of content, often pushes ethics formally out of courses and into circumstances that assume ethics to be 'threaded' or 'woven' across a professional program. As such, the inherent value of ethics can be lost, along with creative means of collaboration. Students need dedicated health care ethics courses with real life examples and not just clear cut, simple case studies with objective ethics principles and theory. Students need to find/see/locate themselves in the learning experience and integrate ethics into who they are and how they might need it for future practice as a health care provider. Upon graduation and out in the real world, academic ideals and perfect professional standards will be challenged and this will require (but may not be met with) a sophisticated level of ethical navigation (Kelly 1998). How can this be achieved?

An arts-informed approach to education is one pathway, with many offshoots, to achieve this end (see Table 10.1 for examples). An innovative Australian example of creative ethics education was offered at Queensland University of Technology over a decade ago (Milligan and Woodley 2009). The course provided a moral space using a patient vignette and culminated in the creation of songs, poetry/prose, and artwork to express a relational understanding of ethics in practice. Faculty encouraged first year paramedic students to view the circumstances of 'patient-other' through four unique lenses (i.e., transformative, hermeneutical, appraisive, and appreciative) and then engaged them in a creative expressive task. After establishing trust (to share a journey of non-conventional learning), this pedagogical approach allowed students to develop not only knowledge and skills but also deep empathy and capacity for ethical sensitivity.

Table 10.1 Arts-informed educational approaches[a]

Types	Examples	Studies
Visual	Collages	Schwind et al. (2014a, b)
	Digital stories	LeBlanc (2017)
	Drawing	Lyon et al. (2013), Rabow et al. (2013), Walji-Jivraj and Schwind (2017)
	Handmade dolls	Demir (2019)
	Inner-outer boxes	Schreiner and Wolf Boronaro (2012)
	Mandalas	Mahar et al. (2012)
	Mask-making	Lordly (2014), Spadoni et al. (2015)
	Multimedia storytelling	Viscardis et al. (2019)
	Painting	Thompson et al. (2010), White et al. (2010)
	Pecha Kucha presentations	Filice and Dampier (2018)
	Photography	Miller et al. (2014)
	Posters	Wehbi (2014)
	Sculpture	Lordly (2014)
Written	Literature	Kinsella (2007), Weisberg and Duffin (1995)
	Prose/poetry	Caeiro et al. (2014), Collett and McLachlan (2006); Connor-Greene et al. (2005), Milligan and Woodley (2009)
	Zines	Desyllas and Sinclair (2014)

[a]Inspired by Kinsella and Bidinosti's (2016), further examples added

A more recent example of an arts-based approach to health care ethics education was offered to occupational therapy students in a graduate course at Western University in Ontario, Canada (Kinsella and Bidinosti 2016). Inspired by arts-informed research in medicine (such as Kumagai and Wear 2014), nursing (such as Schwind et al. 2014a), physical therapy (such as Caeiro et al. 2014), allied health (such as Blomqvist et al. 2007), social work (such as Desyllas and Sinclair 2014), dietetics (such as Lordly 2014), psychology (such as Connor-Greene et al. 2005), and health leadership (such as Hughes 2011), researchers found key themes related to students' discovery of their own sense of: values, creativity, reflective practice, self-awareness, and imaging future self/practice. Further, engaging students with a personalized approach to ethics knowledge and skills may foster humanistic and compassionate practitioners (Kinsella and Bidinosti 2016).

An example of an interprofessional, arts-informed approach to undergraduate ethics education was offered at Queen's University in Ontario, Canada, over 25 years ago (Weisberg and Duffin 1995). The pedagogy and outcomes of this course appear to remain quite relevant today. Although not presented in the context of being arts-informed, the course used various published stories to study professional ethics, professional cultures, and professionalism with nursing, medicine, and law students. Faculty used pedagogy such as freewriting (e.g., journals), images, formal written assignments, and in-class discussions (listening to others) to explore key themes of healing, death, power, and choice. The course allowed students to connect who they are to their professional role, explore value differences, create alliances, challenge stereotypes and prejudices, and tolerate ethical ambiguity.

Finally, an innovative example of internationally collaborative ethics education crossed the ocean between the United States and Britain (Leppa and Terry 2004). Nursing students were exposed to in-class sessions, on-line interactions, and international exchanges of faculty and students to explore and reflect cross-culturally on the ethical meaning of practice. Using real clinical experiences, the students learned about new ways to communicate with each other, discussed alternative options to manage ethics in practice, and had their assumptions challenged.

10.4 Ongoing Individual Capacity Building

After graduation the learning must continue for health care professionals (throughout their career) as novel experiences in practice challenge integrating new knowledge into fixed systems of belief and understanding. Building capacity for the moral self of each individual to act, not just rationally but also compassionately, allows health care providers to maintain their personal and professional integrity and to remain whole instead of becoming shattered. In a Canadian study with 87 nurse participants, researchers held focus groups to explore nurses' human involvement in ethical nursing practice (Doane et al. 2004). Findings revealed: (1) a need to reconcile the gap between personal and professional (self as an individual versus self as a nurse); (2) struggles to navigate role expectations and boundaries (nurse self versus context of practice); and, (3) opportunities to find moral voice, moral imagination, and moral agency through ongoing ethics education. Implications of this study indicate that nurses can build ethical capacity through knowledge and skill development, time to engage and reflect, and with adequate moral support. The recipe for building individual health care providers' ethical capacity is surprisingly simple and is echoed across many academic studies. Still these strategies remains difficult to implement and even harder to sustain due to a constellation of factors.

First, nurses require the time and space to contemplate, reflect back on, and/or explore moral aspects of their practice. As noted in previous chapters, if there is one thing in health care that there is never enough of, it is time. For nurses in Ontario, Canada, reflective practice is a professional obligation (College of Nurses of Ontario 2015). Finding the time to engage and reflect can be one challenge. Another challenge is actually wanting to retell and/or relive difficult moral experiences. Making the time can be hindered by many other pressing obligations and also uncertainties about how to reflect (process) and what the point would be beyond adding to one's daily diet of exposure to suffering. Therefore, the strength of initial strategies that are guided (or facilitated by others) is likely an important factor for success.

In a Canadian study, busyness among nurses was explored in the context of time as a barrier to accessing research for evidence-informed practice (Thompson et al. 2008). Researchers defined 'busyness' as an individual's perception of pressure created by a limited amount of time to do valued work. In this study, influences on nurses' busyness were categorized as being environmental, interpersonal, intrapersonal, and cultural (e.g., nursing culture that values tasks). Also, busyness was connected with heavy workloads and nurses' valuing the provision of satisfactory care to their patients. So, there are valid reasons why nurses and other health care providers are very busy and have limited time to spare.

Further, knowledge and skills must be built up in the area of being able to tolerate failure, because personally achieving exclusively successful ethical decision making is a not realistic expectation. And, if someone cannot tolerate experiencing failure, they are unlikely to choose to review it and accept valuable lessons from that experience. In fact, they will likely avoid or deny it as much as possible. In western culture there is a prioritization on individual achievement (rather than collective progress). American psychology researcher, Angela Duckworth (2016), refers to many high achievers as 'fragile perfects' who only know how to be successful, but do not know what to do with failure. Therefore failure can be a breaking point for some individuals. Her research has identified a lack of growth mindset, resilience, and 'grit'. Further, Duckworth's work with individuals who have 'grit' has correlated it with greater life satisfaction, healthy emotions, and overall wellbeing.

In other words, there is a two-fold reason to develop or enhance your grit or 'sisu.' It can assist with improvements in professional practice as well as individual health, wellbeing, and satisfaction. '*Sisu*' is a Finnish concept that indicates embodied fortitude and bold tenacity. In a study exploring the concept of sisu, key aspects were found to be extraordinary perseverance (with high self-accountability), an action mindset (with consistent courage), and latent power (with visceral and somatic qualities) (Lahti 2019). Important to note, this research identified that sisu can also have destructive qualities, such as harm to self and/or others and reducing the ability to reason effectively. Sisu can be a benefit or a burden.

Knowledge and skills development in mindfulness can also be helpful in many ways. Exploring mindfulness can develop capacity for managing restlessness, finding awareness in the present moment, gaining acceptance and curiosity (rather than judgment), and receiving unexpected healing (Boyce 2011). While eastern religious traditions have known this for centuries, for some reason many in western societies scoff at the profound potential of such seemingly simple practices. Former American army captain turned Ph.D. scholar in mindfulness, Elizabeth Stanley, has the research to back up these important claims about mindfulness. Her research with Jha and colleagues has linked mindfulness with improved working memory capacity and better emotional regulation (Jha et al. 2010), and increased attention to task and cognitive resilience (Jha et al. 2017) among high-stress military cohorts.

But has this approach been used in health care? Yes it has. As an example, a mindfulness based stress reduction (MBSR) approach was used as an educational intervention among medical students in a 10-week seminar and found statistically significant positive effects on their overall psychological distress and mood (Rosenzweig et al. 2003). Also, MBSR was used in a study with oncology nurses as a quality improvement initiative and researchers' found that the training decreased the frequency of their moral distress experiences (Vaclavik et al. 2018).

Finally, a basic approach to mitigating moral distress and supporting ethical decision making takes into consideration the amount and quality of sleep that health care providers are able to obtain. Nurses and other health care providers who work a variety of different shifts often struggle to get enough sleep (not enough time), to get to sleep or stay asleep (insomnia), or to have a high quality of sleep (waking feeling refreshed). In an Italian study, nurses who worked day shift were compared to nurses who worked nights in the context of their psychological and physical health (Ferri et al. 2016). From the perspective of 213 nurses in this study (across 17 hospital units), it was found that night shift work in particular carries a high risk for poor mental and physical health effects, and is linked with job dissatisfaction. However, in contrast an Australian review study of 37 articles on nurses, shift work, psychological functioning, and resilience, suggested that the impacts of shiftwork are largely dependent on contextual and individual factors rather than just the circumstances of shiftwork itself (Tahghighi et al. 2017).

Further, in a study conducted in Israel, researchers found that physicians who worked 24-hour shifts displayed negative impacts on decision making and psychomotor skills abilities (Aran et al. 2017). Physicians who were able to take a nap faired better by making less risky decisions and were able to have more sustained attention. So, there is an argument for attention to sleep and building on other personal skills and resources. Additional strategies for individual capacity building strategies include regular physical exercise, mind–body training (such as yoga), cognitive restructuring/reframing, radical acceptance, learned optimism, meditation, spiritual practices, and gratitude. Additionally, health care providers may seek to learn more about their rights in the workplace and ways to advocate through unions, professional associations, and/or labour boards to ensure that their fundamental rights are upheld and supported (Ross et al. 2018).

Many of the strategies suggested for individual capacity building require an initial stance of self-valuation. Self-valuation requires a growth mindset that prioritizes personal health and wellbeing and accepts imperfection. This is the basis for self-care. It has been my experience that any mention of self-care to an audience of health care professionals (or students) often and almost instantaneously emits a cue for eye-rolls and chuffing. Back to the keen arguments for lack of time, space, energy, motivation, etc. If not to begin with the tactical, then we must discuss the practical. What to do in the moment?

> **Recap of Concept: Individual Ethics Capacity Building**
>
> ✓ Address the gap: personal versus professional
> ✓ Understand role expectations and boundaries
> ✓ Engage in ongoing, meaningful ethics education
> ✓ Reflect and learn from failures
> ✓ Explore mindfulness skills
> ✓ Celebrate your grit or sisu
> ✓ Value self-care…SLEEP!

American professor of business law and ethics, Joshua Perry (2011), suggests five steps to guide ethical decision making and to manage moral distress. See Fig. 10.1. The first three steps include recognizing an ethical issue (versus having ethical blindness), seeking a more full understanding and articulation of the situation, and discerning the possibilities for action (brainstorming). The last two steps involve critical analysis, initially of the options and later of the chosen option and its rationale. Perry suggests an iterative process where, after some time, the ethical issues and process are reflected on in the context of individual and organizational needs for change.

10.5 Team Development and Collaborative Models of Practice

Time to engage and reflect on ethics in practice can happen individually or in a group. If in a group environment, the need to establish a psychologically safe space to share is paramount. As one example, Clarian Health in Indianapolis, Indiana established facilitated ethics conversations for groups and individual walking rounds for dyads to allow staff to share difficult moral experiences and to increase confidence with ethics in practice (Wocial et al. 2010). Various strategies were used to stimulate discussion including probing for further details, guiding articulation of problems, *'what would you do'* polling, summarizing with supportive statements, seeking choices (versus

Fig. 10.1 5 Steps for addressing moral matters

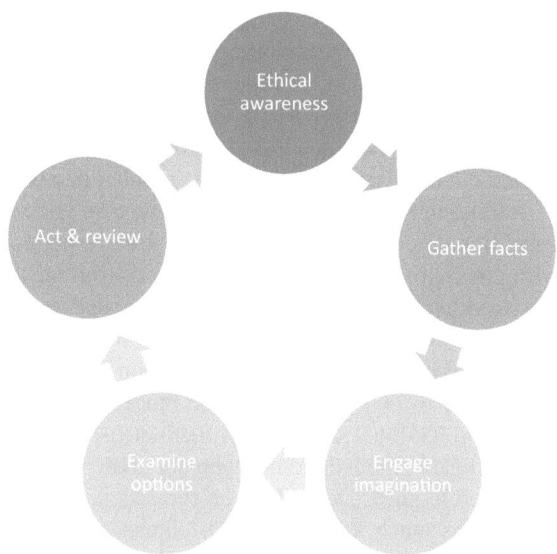

solutions), and exploring best practices (Helft et al. 2009). There are many ways to incorporate ethics discussions into everyday practice activities such as team debriefings, huddles, rounds, and/or as a recurring item on staff meeting agendas. These strategies can be an effective source of moral support that are often overlooked.

In an American study, researchers explored reasons why nurses wanted to build their ethics capacity and participate in a clinical ethics residence program (Jurchak et al. 2017). Nurses spoke of their need to better navigate the 'grey zones' of complex patient situations by gaining ethics knowledge and skills and to guiding patient-centred ethical decision making. However, under many circumstances ethics in practice is addressed as an acute, emergency situation. Ethics is not on the radar until it is needed 'stat'. A moral circumstance that may have been brewing (perhaps unnoticed) suddenly requires real time response and is more likely to have developed into a highly complex and sometimes highly emotionally-charged issue. This is another area where team ethics capacity can be helpful, as ethics committee members (who are usually volunteering their time to the role) are not always immediately available.

Further, the unique role of a nurse ethicist can bridge ethics needs for clinical services consultations, research, and education in health care practice (Wocial et al. 2010). The work of the nurse ethicist can also support ongoing training of individuals to build capacity for organizational ethics champions within a community of practice so that one person (e.g., a nurse ethicist) does not seem to be the only resource.

As an example of a focused effort toward enhancing group ethics knowledge and skill development, a Cincinnati, Ohio hospital held paid workshops (2 hours a week × 4 weeks) for 25 registered nursing staff from their intensive care unit (ICU) department with the hope of decreasing individual levels of moral distress (Beumer 2008). The workshop was led by a nurse manager, clinical nurse specialist, and an employee assistance counsellor. It involved education about moral distress and self-care activities/resources, sharing distressing patient care experiences, developing a personal action plan, and identifying unit specific actions. The workshops were successful in that they showed a decrease in levels of moral distress and the nurses reported feeling validated in their experiences. However, no further workshops were carried out due to time factors as the number one barrier.

Additionally, models of care and how care is delivered should be considered in terms of their contributions to moral distress experiences and as barriers to engaging with ethics in practice. If time and workload are key factors, then how can time and workload be streamlined and/or allocated more appropriately? This is in the interest of working smarter, not harder. In an American study of over two dozen high-functioning organizations, researchers found that a shift way from a physician-centric approach and toward a shared-care model contributed to better team functioning, greater professional satisfaction, and increased personal joy (Sinsky et al. 2013).

And finally, the individuals who are invited to and included in ethics discussions may need to shift. Depending on the sector of health care, non-traditional health professions (such as lawyers; Castellanos and Gillis 2019) and chaplains (Carey et al. 2016) can potentially bring new perspectives, skills, and resources to the table. Further, the role of patients and families is rapidly changing. Not just a voice at the bedside, patient and family representatives now sit on many organizational boards and advisory groups to bring valuable lived experience and first hand knowledge to decisions and policy making.

10.6 Organizational Commitments

When considering the various opportunities that might be available to acknowledge and address moral distress experiences in an organization, it is important to recognize that organizational priorities may be different from the priorities of individual health care providers or teams. Senior leaders generally speak in terms of efficiencies, cost reduction, outcome attainment, and bottom lines on budgets. This is an important factor in terms of understanding why moral distress matters. Addressing moral distress makes sense and it also saves money (cents). Additionally, when and how leaders respond to moral distress can make a difference between hope and hopelessness for change (Pauly et al. 2012). Despite different priorities, common moral commitments must be considered. Everyone, including the organization itself, is supposed to be prioritizing quality patient care. Addressing moral distress and supporting ethics in practice contributes to attaining that shared goal.

One approach to understanding ethics in an organization is to explore and address ethical climate. As previously mentioned, ethical climate can be defined as the practices, conditions, and character of an organization in the management of ethical issues, and has been found to be inversely correlated with moral distress (Pauly et al. 2009). An organization's ethical climate can be assessed using Olson's Hospital Ethical Climate Survey (HECS), although to be fair the organization is only one part of the areas assessed (namely, hospital, managers, peers, patients, and physicians) (Olson 1998).

In terms of an organization's ethical climate, a mixed-methods study of over 200 American nurses revealed that inadequate staffing and incompetence among coworkers were the main contributors to moral distress in practice (Sauerland et al. 2014). Almost ¼ of participants in the study had previously left a position and perceived their current ethical climate more negatively due to past moral distress experiences. Staffing, staff training, employee retention and attrition, and other aspects of ethical climate are within the purview of the organization to address.

Sometimes organizations make an honest effort to address moral distress (or compassion fatigue, burnout, etc.) and to boost ethical climate, but select a strategy that doesn't hit home and devalues the complexity and critical impacts of the symptoms of these experiences. In such a case, a haphazard effort can do more harm than good. Jillian Horton, a physician working in Winnipeg, Manitoba, Canada recalls participating in an organizational 'wellness day' that was simply a plate of muffins (Horton 2020).

Serious and ongoing organizational commitments to promoting a culture of health and wellbeing are needed. In a Canadian study of moral distress and professional quality of life among nurses and physicians, high levels of moral distress and low levels of professional quality of life were found in those experiencing a high load of critically ill patients and among those compelled to provide non-beneficial care (Austin et al. 2017). The Professional Quality of Life (ProQOL; Stamm 2009) scale was used, which has sections to measure both compassion fatigue (including burnout and secondary traumatic stress) and compassion satisfaction.

Also, leadership of organizations has been found to affect wellbeing and professional satisfaction among employees. In an American study at the Mayo clinic, researchers explored burnout, satisfaction, and leadership qualities of supervisors among physicians and found a direct effect (Shanafelt et al. 2015). Therefore, further to supporting staff and efforts toward a positive ethical climate, organizations can also focus their attention on supporting members of their senior leadership teams, supervisors, and managers.

10.7   Knowledge Power Paradox

Another paradox of ethics in practice is that individuals with the knowledge, skills, and/or the vantage point to see and do something about a moral issue may not possess the power to speak up, to be heard, or to effect action. Further, those with the power to address moral issues in practice may not possess the needed knowledge, skills, or perspective(s). This paradox speaks to reasoning behind why ethics needs to be seen as everyone's job (even if there is a bioethicist or ethics champion on staff) and why ethics in practice needs to be a collaborative effort.

In a Canadian study of Advanced Practice Nurses (APNs) and power, eight nurses spoke about their perceptions of building power (or power creep) and sharing power with others (Schoales et al. 2018). Using the term 'soft power' (Nye 2004), researchers described how the APNs used 'power with' (instead of 'power over') to achieve shared goals and acknowledge the value of understanding relationships and context in order to make a difference in their nursing practice.

Advocacy for moral issues and ethical responses in practice is required from many levels, from the highest levels of senior leadership and boards of directors to frontline and behind-the-scenes staff members. Many health care organizations have formally adopted ethical decision making frameworks to guide their employees and supervisors. This strategy is only as effective as the number of individuals who are aware of the framework, know how to access and employ it, and actually use it to achieve moral outcomes in their practice. Otherwise the purpose of an organizational ethics framework may be little more than to achieve a standard of practice on paper; check the box for having established a framework (regardless of its use or usefulness).

Further organizational policies and procedures need to uphold and support quality ethical practice and not impede valuable moral judgement calls and appropriate action.

10.8   Recognizing the Need for Action

In a large American study of 3,000 nurses and social workers from four different states, the preferred response to a moral issue was to talk to a colleague, and nurses were found to be more likely than social workers to not take any action to address the issue (Grady et al. 2008). Surprisingly this is not a unique finding. Other studies have also found that nurses' responses to experiences of moral distress and ethical concerns often lead to no further action (Erlen and Frost 1991; Penticuff and Walden 2000; Sleutel 2000). These consistent findings speak to the very real risks and consequences that nurses and other health care providers can and do face. I would argue that 'doing nothing' is not likely their first choice, but rather a last resort.

American political science scholar, Joan Tronto, has written extensively on the ethics of care and wisely offers insight into some of the tensions in contemporary health care environments that create opportunities for 'good care' to occur. She urges consideration for the 3Ps of purpose, particularity, and power (Tronto 2010). Purpose relates to organizations and individuals and their attention to the shared outcomes of care processes. Like the concept of synergy, the whole is equal to more than the sum of the parts. She also includes the needs of care providers, and not just patients, as an important aspect of achieving desired (moral) outcomes. Particularity relates to meeting the specific, and often fluid, real needs of the care receivers, as determined by both the care receivers and those responsible for care giving. And finally, but definitely not last, she speaks to the influence of power. She draws attention to the misguided concept of patients as customers or consumers who need to be satisfied in a system rife with scarcity of the commodity of care, rather than caring as a fulfilling human activity provided to other humans (who require caring, as all humans do). Tronto suggests the need to 'consciously create' dialogue to recognize and respond to the needs of those giving and those receiving care to determine valuable next steps and actions.

10.9 Conclusion

Research on moral distress by American nurse scholar Ann Hamric (2012) recommends these next steps for moral distress research: further conducting multi-site and multi-disciplinary studies (with findings specific to type of care provider), replication studies with established measures of moral distress, measuring moral distress experiences over time, building a foundation of root causes, and exploring the effects of moral distress on patient/family experiences and outcomes.

Canadian nurse scholar and educator Janet Rankin demands a call to action; for nurses to be political in their efforts to bring forward critical conversations (Rankin 2009). She has written extensively on hospital restructuring and impacts on nursing practice and ultimately patient care outcomes. Her work speaks directly to organizations in the language of factors that truly have an impact on sustainability of the nursing workforce in terms of capacity for recruitment and retention. Rankin wisely refers to the power of the critical mass as a key strategy for nurses to be able to reclaim their professional practice and address barriers to quality nursing care. Hers is a call to action not only for nurses, but for all health care providers.

In conclusion, it is reasonable to admit that we would never expect the confident execution of a complex surgical procedure the first time a physician picks up a scalpel in an emergency situation with a patient. Why then do we have this expectation with ethics in practice? This does not need to remain the status quo. There are both tactical and practical options that can be pursued. While positive change may, at times, seem impossible, adversity and excellence often go hand in hand (Rosenthal 2013).

Being a health care provider and doing the self-sacrificing moral work of caring has an inherent worth because it 'pays dividends' to others (Duckworth 2016). Everyday contributions to alleviate the suffering of self and others 'makes a difference.' This notion of wanting to make a difference is often a catalyst for a career in health care in the first place. I know it was, and still is, for me.

Thus, moral distress experiences can be keenly utilized as a valuable asset to our personal and professional moral global positioning systems (GPS) to help with navigation through the messiness of contemporary health care practice. While codes of ethics and standards of professional practice may set the longitude and latitude, it is the individual health care provider who must make the daily moral decisions about their work with others and their own sense of self and adjust the course accordingly. Further, opportunities are required for ongoing ethics knowledge and skills development, time to reflect and heal (asking for help as needed), and moral support with others along the way.

References

Aran, A., N. Wasserteil, I. Gross, J. Mendlovic, and Y. Pollak. 2017. Medical decisions of pediatric residents turn riskier after a 24 hour call with no sleep. *Medical Decision Making* 37 (1): 127–133. https://doi.org/10.1177/0272989X15626398.

Austin, C.L., R. Saylor, and P.J. Finley. 2017. Moral distress in physicians and nurses: Impact on professional quality of life and turnover. *Psychological Trauma: Theory, Research, Practice, and Policy* 9 (4): 399–406.

Beumer, C.M. 2008. Innovative solutions: The effect of a workshop on reducing the experience of moral distress in an intensive care unit setting. *Dimensions of Critical Care Nursing* 27 (6): 263–267. https://doi.org/10.1097/01.DCC.0000338871.77658.03.

Blomqvist, L., K. Pitkala, and P. Routasalo. 2007. Images of loneliness: Using art as an educational method in professional training. *The Journal of Continuing Education in Nursing* 38 (2): 89–93.

Boyce, B. 2011. *The mindfulness revolution.* Boston, Massachussetts: Shambala Publications.

Caeiro, C., E.B. Cruz, and C.M. Pereira. 2014. Arts, literature and reflective writing as educational strategies to promote narrative reasoning capabilities among physiotherapy students. *Physiotherapy Theory and Practice.* https://doi.org/10.3109/09593985.2014.928919.

Campbell, S.M., C.M. Ulrich, and C. Grady. 2016. A broader understanding of moral distress. *American Journal of Bioethics* 16: 2–9. https://doi.org/10.1080/15265161.2016.1239782.

Carey, L.B., T.J. Hodgson, L. Krikheli, R.Y. Soh, A.R. Armour, T.K. Singh, and C.G. Impiombato. 2016. Moral injury, spiritual care and the role of chaplains: An exploratory scoping review of literature and resources. *Journal of Religion and Health* 55 (4): 1218–1245. https://doi.org/10.1007/s10943-016-0231-x.

Carse, A., and C.H. Rushton. 2017. Harnessing the promise of moral distress: A call for re-orientation. *Journal of Clinical Ethics* 28 (1): 15–29.

Castellanos, N., and M. Gillis. 2019. The lawyer on your side: The power of the interprofessional team in preventing moral distress and empathy erosion. In *Teaching empathy in healthcare: Building a new core competency*, ed. A. Foster and Z. Yaseen, 269–284. Netherlands: Springer International Publishing.

College of Nurses of Ontario. 2015. *Practice reflection: Learning from practice.* Toronto, ON: Author. Retrieved from https://www.cno.org/globalassets/4-learnaboutstandardsandguidelines/prac/learn/teleconferences/practice-reflection--learning-from-practice.pdf

Collett, T.J., and J.C. McLachlan. 2006. Evaluating a poetry workshop in medical education. *Medical Humanities* 32 (1): 59–64. https://doi.org/10.1136/jmh.2005.000222.

Connor-Greene, P.A., A. Young, C. Paul, and J.W. Murdoch. 2005. Poetry: It's not just for English class anymore. *Teaching of Psychology* 32 (4): 215–221. https://doi.org/10.1207/s15328023top3204_2.

Demir, E. 2019. An analysis of nursing students' experiences about handmade nursing dolls during nursing education. *International Journal of Caring Sciences* 12 (1): 286–292.

Desyllas, M.C., and A. Sinclair. 2014. Zine-making as a pedagogical tool for transformative learning in social work education. *Social Work Education* 33 (3): 296–316. https://doi.org/10.1080/02615479.2013.805194.

Doane, G., B. Pauly, H. Brown, H., and G. McPherson, G. 2004. Exploring the heart of ethical nursing practice: implications for ethics education. *Nursing Ethics* 11 (3): 240–253. https://doi.org/10.1191/0969733004ne692oa.

Duckworth, A. 2016. *Grit: The power of passion and perseverance.* Toronto, ON: HarperCollins.

Erlen, J.A., and B. Frost. 1991. Nurses' perceptions of powerlessness in influencing ethical decisions. *Western Journal of Nursing Research* 13 (3): 397–407.

Ferri, P., M. Guadi, L. Marcheselli, S. Balduzzi, D. Magnani, and R. Di Lorenzo. 2016. The impact of shift work on the psychological and physical health of nurses in a general hospital: A comparison between rotating night shifts and day shifts. *Risk Management and Healthcare Policy* 9: 203–211. https://doi.org/10.2147/RMHP.S115326.

Filice, S., and S. Dampier. 2018. Experiential community health assessment through Pecha Kucha. *Journal of Nursing Education* 57 (9): 566–569. https://doi.org/10.3928/01484834-20180815-11.

Grady, C., M. Danis, K.L. Soeken, P. O'Donnell, C. Taylor, A. Farrar, and C.M. Ulrich. 2008. Does ethics education influence the moral action of practicing nurses and social workers? *The American Journal of Bioethics* 8 (4): 4–11. https://doi.org/10.1080/15265160802166017.

Hamric, A.B. 2012. Empirical research on moral distress: Issues, challenges, and opportunities. *HEC Forum* 241: 39–49. https://doi.org/10.1007/s10730-012-9177-x.

Helft, P.R., P.D. Bledsoe, M. Hancock, and L.D. Wocial. 2009. Facilitated ethics conversations: A novel program for managing moral distress in bedside nursing staff. *JONA'S Healthcare Law, Ethics and Regulation* 11 (1): 27–33.

Horton, J. 2020. Your burnout isn't your fault, but you should 'own' it. *Medscape*. Retrieved from https://www.medscape.com/viewarticle/923610_1.

Hughes, S. 2011. The leadership mask: A personally focused art based learning inquiry into facets of leadership. *Reflective Practice* 12 (3): 305–331. https://doi.org/10.1080/14623943.2011.571863.

Jha, A.P., A.B. Morrison, S.C. Parker, and E.A. Stanley. 2017. Practice is protective: Mindfulness training promotes cognitive resilience in high-stress cohorts. *Mindfulness* 8 (1): 46–58. https://doi.org/10.1007/s12671-015-0465-9.

Jha, A.P., E.A. Stanley, A. Kiyonaga, L. Wong, and L. Gelfand. 2010. Examining the protective effects of mindfulness training on working memory capacity and affective experience. *Emotion* 10 (1): 54–64. https://doi.org/10.1037/a0018438.

Jurchak, M., P.J. Grace, S.M. Lee, D.G. Willis, A.A. Zollfrank, and E.M. Robinson. 2017. Developing abilities to navigate through the grey zones in complex environments: Nurses' reasons for applying to a clinical ethics residency for nurses. *Journal of Nursing Scholarship* 49 (4): 445–455. https://doi.org/10.1111/jnu.12297.

Kälvemark Sporrong, S., B. Arnetz, M.G. Hansson, P. Westerholm, and A.T. Höglund. 2007. Developing ethical competence in health care organizations. *Nursing Ethics* 14 (6): 825–837. https://doi.org/10.1177/0969733007082142.

Kelly, B. 1998. Preserving moral integrity: A follow-up study with new graduate nurses. *Journal of Advanced Nursing* 28: 1134–1145.

Kinsella, E.A. 2007. Educating socially responsive practitioners: What can the literary arts offer health professional education? In *The arts and social justice: Re-crafting adult education and community cultural leadership*, ed. D. Clover and J. Stalker, 39–58. Leicester: National Institute for Adult Continuing Education.

Kinsella, E.A., and S. Bidinosti. 2016. 'I now have a visual image in my mind and it is something I will never forget': An analysis of an arts-informed approach to health professions ethics education. *Advances in Health Sciences Education* 21 (2): 303–322. https://doi.org/10.1007/s10459-015-9628-7.

Kumagai, A.K., and D. Wear. 2014. "Making strange": A role for the humanities in medical education. *Academic Medicine* 89 (7): 973–977. https://doi.org/10.1097/ACM0000000000000269.

Lahti, E.E. 2019. Embodied fortitude: An introduction to the Finnish construct of sisu. *International Journal of Wellbeing* 9 (1): 61–82. https://doi.org/10.5502/ijw.v9i1.672.

LeBlanc, R.G. 2017. Digital story telling in social justice nursing education. *Public Health Nursing* 34 (4): 395–400. https://doi.org/10.1111/phn.12337.

Leppa, C.J., and L.M. Terry. 2004. Reflective practice in nursing ethics education: International collaboration. *Journal of Advanced Nursing* 48 (2): 195–202.

Lordly, D. 2014. Crafting meaning: Arts-informed dietetics education. *Canadian Journal of Dietetic Practice and Research* 75 (2): 89–94. https://doi.org/10.3148/75.2.2014.89.

Lovecky, D.V. 1997. Identity development in gifted children: Moral sensitivity. *Roeper Review* 20 (2): 90–94. https://doi.org/10.1080/02783199709553862.

Lyon, P., P. Letschka, T. Ainsworth, T., and I. Haq. 2013. An exploratory study of the potential learning benefits for medical students in collaborative drawing: Creativity, reflection and 'critical looking'. *BMC Medical Education* 13 (1): 86. https://doi.org/10.1186/1472-6920-13-86.

Mahar, D.J., C.L. Iwasiw, and M.K. Evans. 2012. The mandala: First-year undergraduate nursing students' learning experiences. *International Journal of Nursing Education Scholarship* 5 (9): 26. https://doi.org/10.1515/1548.923X.2313.

McCarthy, J., and R. Deady, R. 2008. Moral distress reconsidered. *Nursing Ethics* 15 (2): 254–262. https://doi.org/10.1177/0969733007086023.

Miller, E., D. Balmer, N. Hermann, G. Graham, and R. Charon. 2014. Sounding narrative medicine: Studying students' professional identity development at Columbia University College of Physicians and Surgeons. *Academic Medicine* 89 (2): 335–342. https://doi.org/10.1097/ACM.000000 0000000098.

Milligan, E., and E. Woodley. 2009. Creative expressive encounters in health ethics education: Teaching ethics as relational engagement. *Teaching and Learning in Medicine* 21 (2): 131–139. https://doi.org/10.1080/10401330902791248.

Nye, J.S. 2004. *Soft power: The means to success in world politics.* New York, NY: Public Affairs.

Olson, L. 1998. Hospital nurses' perceptions of the ethical climate of their work setting. *Journal of Nursing Scholarship* 30 (4): 345–349.

Pauly, B., C. Varcoe, and J. Storch. 2012. Framing the issues: Moral distress in health care. *HEC Forum* 24 (1): 1–11. https://doi.org/10.1007/s10730-012-9176-y.

Pauly, B., C. Varcoe, J. Storch, and L. Newton. 2009. Registered nurses' perceptions of moral distress and ethical climate. *Nursing Ethics* 16 (5): 561–573. https://doi.org/10.1177/096973300 9106649.

Penticuff, J.H., and M. Walden. 2000. Influence of practice environment and nurse characteristics on perinatal nurses' responses to ethical dilemmas. *Nursing Research* 49 (2): 64–72.

Perry, J.E. 2011. Managing moral distress: A strategy for resolving ethical dilemmas. *Business Horizons* 54 (5): 393–397. https://doi.org/10.1016/j.bushor.2011.05.005.

Rabow, M.W., C.N. Evans, and R.N. Remen. 2013. Professional formation and deformation: Repression of personal values and qualities in medical education. *Family Medicine* 45 (1): 14–18. https://doi.org/10.1097/ACM.0b013e3181c887f7.

Rankin, J.M. 2009. The nurse project: An analysis for nurses to take back our work. *Nursing Inquiry* 16 (4): 275–286.

Rogers, S., A. Babgi, and C. Gomez, C. 2008. Educational interventions in end-of-life care: Part I an educational intervention responding to the moral distress of NICU nurses provided by an ethics consultation team. *Advances in Neonatal Care* 8 (1): 56–65. https://doi.org/10.1097/01. ANC.0000311017.02005.20.

Rosenthal, N.E. 2013. *The gift of adversity: The unexpected benefits of life's difficulties, setbacks, and imperfections.* New York, NY: Penguin Books.

Rosenzweig, S., D.K. Reibel, J.M. Greeson, G.C. Brainard, and M. Hojat, M. 2003. Mindfulness-based stress reduction lowers psychological distress in medical students. *Teaching and Learning in Medicine* 15 (2): 88–92. https://doi.org/10.1207/S15328015TLM1502_03.

Ross, C.A., N.S. Berry, V. Smye, and E.M. Goldner. 2018. A critical review of knowledge on nurses with problematic substance use: The need to move from individual blame to awareness of structural factors. *Nursing Inquiry* 25 (2): e12215. https://doi.org/10.1111/nin.12215.

Sauerland, J., K. Marotta, M.A. Peinemann, A. Berndt, and C. Robichaux. 2014. Assessing and addressing moral distress and ethical climate, part 1. *Dimensions of Critical Care Nursing* 33 (4): 234–245. https://doi.org/10.1097/DCC.0000000000000050.

Schoales, C.A., F.F. Bourbonnais, and J. Rashotte, J. 2018. Building to make a difference: Advanced Practice Nurses' experience of power. *Research and Theory for Nursing Practice* 32 (1): 96–116. https://doi.org/10.1891/0000-000Y.32.1.96.

Schreiner, L. and G.P. Wolf Boronaro. 2012. Inner-outer boxes: An arts-based self-reflection experience about death and dying. *Journal of Hospice and Palliative Nursing* 14 (8): 559–562. https://doi.org/10.1097/NJH.0b013e31825ec187.

Schwind, J.K., H. Beanlands, J. Lapum, D. Romaniuk, and S. Fredericks. 2014. Fostering person-centered care among nursing students: Creative pedagogical approaches to developing personal knowing. *Journal of Nursing Education* 56 (6): 343–347. https://doi.org/10.3928/01484834-201 40520-01.

Schwind, J.K., G.M. Lindsay, S. Coffey, D. Morrison, and B. Mildon. 2014. Opening the black-box of person centred care: An arts-informed narrative inquiry into mental health education and practice. *Nursing Education Today* 34 (8): 1167–1171. https://doi.org/10.1016/j.nedt.2014. 04.010.

Seligman, M.E.P. 2011. *Flourish: A visionary new understanding of happiness and well-being.* New York, NY: Free Press.

Shanafelt, T.D., G. Gorringe, R. Menaker, K.A. Storz, D. Reeves, S.J. Buskirk, J.A. Sloan, and S.J. Swensen. 2015. Impact of organizational leadership on physician burnout and satisfaction. *Mayo Clinic Proceedings* 90 (4): 432–440.

Sinsky, C.A., R. Willard-Grace, A.M. Schutzbank, T.A. Sinsky, D. Margolius, and T. Bodenheimer. 2013. In search of joy in practice: A report of 23 high-functioning primary care practices. *The Annals of Family Medicine* 11 (3): 272–278. https://doi.org/10.1370/afm.1531.

Sleutel, M.R. 2000. Intrapartum nursing care: A case study of supportive interventions and ethical conflicts. *Birth* 27 (1): 38–45.

Spadoni, M., G.H. Doane, P. Sevean, and K. Poole. 2015. First-year nursing students—Developing relational caring practice through inquiry. *Journal of Nursing Education* 54 (5): 270–275. https:// doi.org/10.3928/01484834-20150417-04.

Stamm, B.H. 2009. Professional quality of life: Compassion satisfaction and fatigue version 5 (ProQOL). Retrieved from www.isu.edu/~bhstamm or www.proqol.org.

Tahghighi, M., C.S. Rees, J.A. Brown, L.J. Breen, and D. Hegney. 2017. What is the impact of shift work on the psychological functioning and resilience of nurses? An integrative review. *Journal of Advanced Nursing* 73 (9): 2065–2083.

Thompson, T., C. Loamont-Robinson, and L. Younie, L. 2010. 'Compulsory creativity': Rationales, recipes, and results in the placement of mandatory creative endeavor in a medical undergraduate curriculum. *Medical Education Online* 15. https://doi.org/10.3402/meo.v15i0.5394.

Thompson, D.S., K. O'Leary, E. Jensen, S. Scott-Findlay, L. O'Brien-Pallas, and C.A. Estabrooks. 2008. The relationship between busyness and research utilization: It is about time. *Journal of Clinical Nursing* 17 (4): 539–548. https://doi.org/10.1111/j.1365-2702.2007.01981.x.

Tronto, J.C. 2010. Creating caring institutions: Politics, plurality, and purpose. *Ethics and Social Welfare* 4 (2): 158–171. https://doi.org/10.1080/1749635.2010.484259.

Vaclavik, E.A., B.A. Staffileno, and E. Carlson. 2018. Moral distress: Using mindfulness-based stress reduction interventions to decrease nurse perceptions of distress. *Clinical Journal of Oncology Nursing* 22 (3): 326–332. https://doi.org/10.1188/18.CJON.326-332.

Viscardis, K., C. Rice, V. Pileggi, A. Underhill, E. Chandler, N. Changfoot, P. Montgomery, and R. Mykitiuk. 2019. Difference within and without: Health care providers' engagement with disability arts. *Qualitative Health Research* 29 (9): 1287–1298. https://doi.org/10.1177/104973 2318808252.

Walji-Jivraj, N., and J.K. Schwind. 2017. Nurses' experience of creating an artistic instrument as a form of professional development: an arts-informed narrative inquiry. *International Practice Development Journal* 7 (1). https://doi.org/10/19043/ipdj.71.003.

Wehbi, S. 2014. Arts-informed teaching practice: Examples from a graduate anti-oppression class-room. *Social Work Education* 34 (1): 46–59. https://doi.org/10.1080/02615479.2014.937417.

Weisberg, M., and J. Duffin. 1995. Evoking the moral imagination: Using stories to teach ethics and professionalism to nursing, medical, and law students. *Journal of Medical Humanities* 16 (4): 247–263.

White, C.B., R.L. Perlman, J.C. Fantone, and A.K. Kumagai. 2010. The interpretive project: A creative educational approach to fostering medical students' reflections and advancing humanistic medicine. *Reflective Practice* 11 (4): 517–527. https://doi.org/10.1080/14623943.2010.505718.

Wocial, L.D., P. Bledsoe, P.R. Helft, and L.Q. Everett. 2010. Nurse ethicist: Innovative resource for nurses. *Journal of Professional Nursing* 26 (5): 287–292. https://doi.org/10.1016/j.profnurs. 2010.06.003.

Index

© Springer Nature Switzerland AG 2020
K. Jones-Bonofiglio, *Health Care Ethics through the Lens of Moral Distress*,
The International Library of Bioethics 82,
https://doi.org/10.1007/978-3-030-56156-7